DARWIN'S SECRET SEX PROBLEM

Exposing Evolution's Fatal Flaw—
The Origin of Sex

F. LaGard Smith

WestBow Press books may be ordered through booksellers or by contacting:

WestBow Press
A Division of Thomas Nelson & Zondervan
1663 Liberty Drive
Bloomington, IN 47403
www.westbowpress.com
1 (866) 928-1240

ISBN: 978-1-9736-1706-8 (sc)
ISBN: 978-1-9736-1707-5 (hc)
ISBN: 978-1-9736-1705-1 (e)

Library of Congress Control Number: 2018901379

Printed in the United States of America.

WestBow Press rev. date: 05/03/2018

To tomorrow's researchers, thinkers, teachers, and leaders,
whose passionate search for scientific truth may at last bring about
a long-overdue rejection of the romanticized Evolution Story.

An Evolution Revolution is coming!

ACKNOWLEDGEMENTS

One is left with the feeling that some essential feature of the situation is being overlooked.

—J. Maynard Smith

The thesis of this book is that an essential feature of Darwinian evolution—the origin of sex—is being overlooked, a feature which fatally undermines today's orthodox scientific paradigm of a natural progression from microbe to man. Lest I make the same mistake, I don't want to overlook an essential feature of the evolution of this very book—namely, those individuals who contributed significantly to the volume you hold in your hand.

It was Jed Macosko's out-of-the-blue phone call from his physics lab at Wake Forest that stirred decades of contemplation into action with nothing more than a casual conversation becoming the surprising catalyst to my writing of this book. Through Jed (whose own focus is on the mechanics of protein machines), I also met others whose help has been invaluable along the way, including British-born Geoff Barnard, a former Cambridge researcher and one who, himself, has lectured on the topic of sex and evolution. Geoff's critical eye, all the way from his home in Israel, has kept me firmly on the path.

Jed also led me to Joris Paul van Rossum, a Dutch evolutionist whose writing raised serious doubts regarding natural selection's ability to explain the origin of sex. Despite our coming at the problem from different directions, Joris Paul has stirred my thinking, been a collegial advisor, and contributed in important ways to my analysis in the pages ahead.

There is no way I can adequately thank John Silvius, Senior Professor Emeritus of Biology at Cedarville University, who in the midst of his busy schedule not only did great service in fact-checking my work, but who also helped craft my writing for better clarity, punch, and, hopefully, persuasion. (Undoubtedly, John would have suggested that I break that last sentence into two!)

My appreciation also goes to Phillip Johnson, the author of *Darwin on Trial* and widely regarded as father of the Intelligent Design movement, who read my manuscript from an ID perspective, and was kind enough to endorse a book taking a somewhat different approach. His expressed hope that this book would receive an objective hearing from the scientific community is, of course, my hope as well.

I also wish to thank Frank Breeden, Managing Partner of Premiere Authors, whose faith in the premise of this book and whose insightful suggestions were an encouragement when most needed.

Given sensitive realities on some of their campuses, I want to send an anonymous shout-out to a half-dozen biologists in academia who graciously reviewed my manuscript with an eye to scientific accuracy, and further engaged my arguments regarding theistic evolution.

In an effort to fairly present the ideas of those whose work I've cited (particular those with whom I've taken issue), I have reached out to give an opportunity for review and feedback. I want to deeply thank those who graciously responded. This book is a better book because of their willingness to join in the dialogue.

There are also a number of close friends who were kind enough to read various drafts of this work and share their thoughts. Greater love hath no friend! Among these, I particularly want to thank Nathan Guy, who not only played Devil's Advocate in his own inimitable style, but who opened doors to others whose input was so helpful.

Special thanks to you, Kevin, for your insightful, even adversarial, critique which unquestionably enabled this book to evolve into a more robust species!

While I am grateful for the contributions from all those listed above, I should make it clear on their behalf that none of them is responsible for anything that might be amiss with the book. Indeed, some of them were privy to only parts of the manuscript or to earlier versions. Whatever their role in review and critique, clearly the ideas, arguments, and positions expressed herein are solely my own.

My appreciation also goes to Jack Lyon for his technical assistance in resolving a pesky formatting problem in an early draft, and to Shannon Mohajerin who helped me solve any number of mysteries hidden within my inscrutable Word program.

I also wish to thank Mary Hooper for her suggestions on the cover design. Darwin has never bled so profusely (to the margins)!

And then there's the lovely Ruth, whose creativity and attention to detail provide the wonderfully conducive environment in which I am blessed to do my writing, not the least of which is her enchanting English country garden just outside my window.

CONTENTS

PREFACE

If it could be demonstrated that any complex organ existed, which could not possibly have been formed by numerous, successive, slight modifications, my theory would absolutely break down.
—Charles Darwin

As a reader, you're entitled to ask why someone who's not a science guy would spend several years digging into the science behind the widely-accepted notion of microbe-to-man evolution. In candor, I'd prefer waiting to the end of the book to tell you why I've made this laborious effort, so as to make my case without any fear that it might be dismissed out of hand for wrong reasons. It's important that this book stand on its own merits without being prejudged on the basis of the impetus behind its writing. But because you will be investing time and thought in exploring a rather daunting subject, you deserve at least this much of a heads-up.

While the observable process of natural selection (call it "little-e" evolution) has no adverse philosophical implications whatsoever, the popular microbe-to-man Evolution Story (call it "Big-E" Evolution) has huge philosophical implications for the meaning of human existence, moral values, and (given increasing interest in theistic evolution) even theological and spiritual matters. It would require another book entirely, but—laying my cards squarely on the table—concerns about the wide-ranging ramifications of evolutionary thinking have contributed significantly to my writing this book. Absolutely no two questions are more fundamental than where we came from and why we are here. Every philosophical debate and cultural divide—whether moral, political, economic, or social—eventually traces its way back to the single, pivotal issue of origins.

For those who are enamored with the notion of common descent, never was there a truer version than the origin of all species of *ideas* descending inexorably from a single issue. The evolutionist argument of "population divergence" speaks volumes here, as the question of origins is the original "fork in the road," separating beliefs, ideologies, and basic human values. Which gives us something serious to think about. Because unpurposed, undirected, naturalistic evolution has no moral DNA of its own, if any seemingly-shared moral DNA appears along the path of evolutionary thinking, it could only be a vestigial feature lingering from the persistent influence of the other path. But what if there had never been any other path....?

Too often, and for too long, the battle over origins has been waged between science and religion with neither side seemingly capable of speaking the language of the other. This book is an attempt to bridge that gap by ignoring religious assumptions altogether and, instead, introducing a fresh, powerful argument solely from within the realm of science. Bible or no Bible, Genesis or no Genesis, faith or no faith, Darwin's Grand Theory is fatally flawed on its own terms, as set forth by Darwin himself in the quotation above.

If you are an evolutionist (even theistic evolutionist), I ask only that you hear me out and be willing to consider distinguishing what you rightly accept about natural selection from what is hugely problematic when it comes to the quite separate notion of common descent at the heart of Darwin's Grand Theory. The fundamental problem with the classic Evolution Story is that, scientifically-speaking, the sexual mechanisms required for "Big-E" Evolution could not possibly have happened via the process of "little-e" evolution. What works beautifully on one level doesn't work at all on a higher, overarching level. (Think toy train locomotive being asked to pull a hundred *real* train cars.)

If you are a creationist, I urge you to patiently plow your way (or at least seriously skim) through a rather more intellectually-rigorous argument than you've likely heard before (an argument which, of necessity, speaks explicitly about matters pertaining to sex and sexuality, not to shock or pander, but to persuade).

Creationists may also notice that the approach taken in this book does not fall into the tempting trap laid when evolutionists argue that if "little-e" evolution is scientifically valid, "Big-E" Evolution must be equally valid. Some (few, I think) creationists have foolishly taken the bait of "Big-E" Evolution's classic "bait and switch" argument, thereby wasting undue amounts of time and energy trying to prove that "little-e" evolution (as in natural selection and survival of the fittest) has no basis whatsoever in scientific fact—all in an effort to disprove "Big-E" Evolution for primarily philosophical reasons.

Ironically, this well-intended but misguided strategy has merely served as a recruiting tool for thoughtful scientists who, knowing that "little-e" evolution is true, wrongly assume that "Big-E" Evolution must therefore be logically valid. By contrast, this book does not challenge the plainly obvious mechanisms of "little-e," observable evolution, only the dubious mechanisms of "Big-E," microbe-to-man Evolution—specifically, the mechanisms pertaining to sexual reproduction, both in its origin and along the supposed path of common descent.

My fervent hope is that one day the scientific community will have the courage to repudiate the popular notion of microbe-to-man evolution as grossly-oversold, flawed science. Perhaps then it might finally be possible to seriously address Evolution's ruinous philosophical assumptions that have led us blindly down a wrong path.

But first things first. In the pages ahead, it's not principally about philosophical differences, but about hard science. And about a bedazzling Evolution Story in which bona fide science moves unnoticed from undeniable fact to pure fiction.

Laying Some Important Groundwork

The question is, how can we have this conversation? By its very nature, this book is a polemic, marshalling the strongest arguments I can present in an effort to persuade. But that's not how science thinks. Science is cautious, tentative, content with uncertainty in the pursuit of knowledge. Science probes, speculates, and hypothesizes. That's not an easy audience for a guy like me who's often wrong but never in doubt! Especially since it's *science* I'm writing about.

To be sure, science has its own apologists who are more than happy to lock horns and engage in vigorous, no-holds-barred argumentation, but that merely compounds the difficulty of knowing how to proceed. Who is my primary audience? Is it militant evolutionists like Richard Dawkins and Jerry Coyne (and their ardent fans) for whom the cut and thrust of vigorous debate is standard fare, or is it the hard-working tech in the lab quietly going about his research into intriguing evolutionary processes? Compounding the problem, I suspect that most of my readers will not be scientists at all, but just thoughtful folks interested in hearing a new "take" on the time-worn evolution controversy. So, how do I provide comprehension and readability for the lay person while providing enough gravitas for specialists? Don't mean to moan, just letting you know in advance that my writing style and particular brand of critical thinking may not be every reader's preferred cup of tea, most especially scientists.

Consider, for example, the difference between two similar words: *impossible* and *improbable*. Whereas science is all about the *probable* and the *improbable*, in the pages ahead you'll find me using the word *impossible*, which I appreciate is anathema to a scientist. Might even open me up to the charge of hubris most egregious. I understand. Who am I, of all people, to proclaim that anything in the realm of science is *impossible*? But let's see if we can close the gap a bit. Do you think it's possible for pigs to fly? If you think that's not possible, is it okay to use the word *impossible* rather than just *improbable*?

History, of course, has a way of reminding us that many things once confidently believed to be impossible are now entirely possible. Just ask Wilbur and Orville Wright. "You sayin' people can fly through the sky? Impossible!" "And fly to the moon and back? Impossible!" We skeptical, often-unenlightened humans have collectively embarrassed ourselves too many times to count. Even so, what do you think? Will *pigs* (as pigs) ever fly? Have pigs (as pigs) *ever* flown? Not a chance. Given their very nature, we can all agree, can't we, that it's *impossible* for pigs to fly?

How about this one? Do rocks have sex? To be sure, they can crumble and mix with other crumbled rocks to form new rocks, but, not being living beings, they don't reproduce as sex reproduces. Can we not agree, then, that—because of their very nature—it's *impossible* for rocks to have sex? The point is that, as you read further, I'll be making arguments very much like, "It's impossible for rocks to have sex." There's no hubris in it, and no foolish insistence on the impossibility of something that one day might embarrassingly prove to be clearly possible. *Considering the very nature of things*, you and I can easily agree that it's *impossible* for certain things to happen. Apples won't soon be falling *up* from apple trees instead of down. And snow won't have pink and purple flakes instead of white. Given the inherent nature of things, such phenomena are logically impossible.

Logic, of course, can sometimes fool us. As you'll soon see, I believe that's exactly what happened to Charles Darwin. Based as it was on observable evolution within known species, Darwin's leap to his Grand Theory of evolution from lower to higher species was logical enough… just wrong. Think, for example, about a traffic cop who radars a car going 90 miles an hour. A certain logic could reasonably conclude that, an hour before, the car was 90 miles away. Of course, the far greater likelihood is that the driver pulled the car onto the highway only a few miles back and put the pedal to the metal.

Darwin was not the only one who has fallen victim of the kind of "if/then" reasoning that can be entirely logical, but entirely wrong. On the other hand, to blatantly defy all logic and reason is to, well, defy all logic and reason! So, when we use our logic to conclude that it's *impossible* for pigs to fly, we're logically drawing legitimate conclusions from the known nature of things.

Edging closer to the discussion to follow, what can we logically and legitimately conclude about the outcome of explosions? Do they result in order, or disorder? Has it ever been observed that explosions produce magnificent cathedrals or high-tech flying machines? (Okay, if it explodes in a farm yard, maybe flying pigs!) In the same vein, has it ever been observed that unguided, random forces have produced magnificent cathedrals or high-tech flying machines? If it's order and high functionality we're after, is randomness the first place logic would lead us?

The critical importance of logical thinking remains true even if we might easily *imagine* pigs flying, or apples falling upwards, or pink and purple snowflakes. Yet, here again we begin to talk past each other as scientists and nonscientists. Scientists are all about imagination, taking the form of theories, hypotheses, models, conjecture, educated guesses, and merely wondering "What if?". To our benefit, that unfettered curiosity and sense of sheer wonder has often led to marvelous scientific breakthroughs.

That said, there's a world of difference between the imaginative brainstorming of the scientific researcher and the kind of fanciful imagination that popular evolution writers employ when spinning out the romanticized Evolution Story. Given their penchant for unrestrained imagination (complete with full-color illustrations), much of their story is pure fiction, ripe for unsuspecting, uncritical readers. (If you detect any stridency in my arguments, this kind of gratuitous imagining warrants some serious upbraiding.)

By contrast are those scientists who struggle valiantly with the very issues raised in this book and, in the process, imagine developmental pathways that might have moved the evolutionary ball from Point A to Point B. At times, that imagination (sometimes even whacky imagination!) leads to exciting discoveries. At other times, that imagination can be little more than a futile attempt to explain what can never be explained because of the inherent nature of things.

It's also possible that one doesn't even recognize glaring gaps in scientific logic because their imagination has already been captured by a fascinating, elegant story. When one unreservedly believes that the evolution ball has moved all the way from Point A to Point Z, there is less urgency to explain how it might have gotten from Point A to Point B. That particular gap may be troublesome or highly problematic—even fatal, in this case—but in the end for the true believer it doesn't really matter. Somehow, some way, moving from Point A to Point B on the way to Point Z simply *must* have happened!

At the mention of "true believers," one can hardly help but think of true believers in the realm of religion. We, too, live with uncertainty about many details of our faith which leave us perplexed. We, too, long for explanations we're likely never to have. We, too, at times rationalize what seems all too irrational. We, too, can find ourselves imagining the unimaginable. And why? Because we fervently believe our faith story from A to Z, even if some particular "Point A to Point B" event or "Point C to Point D" doctrine is troublesome and problematic.

Given life's many mysteries that confound us all, absolute certainty is not being demanded of anyone on either side of the aisle, only as much intellectual honesty as we can muster and, to the best of our ability, complete candor. Which leads to this observation....

Whether it be in popular books, textbooks, or student notebooks, the microbe-to-man Evolution Story is no longer presented or understood as legitimate scientific brainstorming, but as undeniable, clearly proven, indisputable scientific fact. Problem is, while speculation *leading* to fact is sound scientific method, mere speculation *presented* as fact is not science, but advocacy—especially when huge gaps remain that, if paid enough attention, would blow the popular story completely out of the water. Which is what this book is all about: insurmountable, unbridgeable gaps.

At this point, I confess a certain boldness to my argument which I urge you to keep in mind as we go along. Far from unfairly demanding immediate, satisfactory, scientific explanations in complete detail as to how crucial gaps in evolution theory might have been bridged (too high a standard for any scientist), what I'm actually saying is that—*given the very nature of things* and the logical conclusions that necessarily follow—*no scientific explanation whatsoever is possible*. "Wow!" I hear you saying. Bold indeed! But in order for the Evolution Story to be true, some truly *impossible* things would have to have happened along the way, defying all logic and reason. Because of those impossible things, pigs (flying or not) would never have existed; nor apples and apple trees; nor, most significantly, *us*.

It's important to note here that no scientist is tempted to say that, conceivably, pigs could fly and apples could fall skyward if only we had enough time and resources to figure out how. Some things are just too obvious. As most scientists will readily acknowledge, some things in Nature truly are *impossible*. So, how are we to distinguish between what science is happy to acknowledge as truly impossible and what science is reluctant to acknowledge as impossible until opportunity for further research and plausible explanation?

Although the line is not as clear as we might wish, the first approach (We know and accept the obvious) readily recognizes the very nature of pigs and falling apples and affirms the logical impossibilities involved. The second approach (We don't yet know what we might some day know) is valid only when impossibility is not demanded by the very nature of whatever is under consideration. The case presented in this book argues that—*given the unique nature of male/female mating and reproduction*—the issues regarding evolutionary sex fall into the former category of obvious impossibility and not into the latter.

Keeping it simple, surely we can all agree that it's *impossible* for a male cheetah in the wild to sire cheetah offspring without a female cheetah, and equally *impossible* for a female cheetah to mate and reproduce cheetah offspring with an elephant. The arguments that lay ahead are but variations on those easily recognized observations. Yet, here's what's curious. If our conversation has nothing at all to do with science or evolution, no one is disturbed in the least by that obvious logic. But, should the discussion turn to evolution, suddenly we start hearing, "Yes, but what if the male cheetah…?" Or, "We don't know how it would be possible for a cheetah and an elephant to reproduce, but we can't definitively rule it out." Or, "There are many studies working on the problem, and one day we might know how that could happen." Why is it that the logic of what's "obviously impossible" suddenly vanishes when talk turns to evolution? With Evolution (taking a page from what's said of God), there is nothing that is *impossible*!

Given such responses from proponents of evolution, does the "obviously impossible" distinction require universal recognition? Must *everyone* agree that it's impossible? Not necessarily. But where there's not the kind of unanimity we have when talking about flying pigs and upwardly-falling apples, the burden of the argument is on anyone claiming "obvious impossibility" in the face of others who disagree (most especially the scientific community). In this book, that heavy burden is clearly mine, and I'm under no illusion but that what's obvious to me (and to many others) may be a hard sell for anyone firmly locked into the Evolution Story.

What I could wish to prove to even the most skeptical reader is that the very nature of sexual reproduction, both in its origin and in its putative role in common descent, forces the logical conclusion that evolutionary sex could not possibly have occurred by natural selection, thereby undermining Darwin's Grand Theory. And thereby also putting paid to the notion that, given enough time and resources, some day we might be able to explain the origin and transmission of sex in the chain of evolution. No matter how many studies we undertake, we're not going to figure out how sex ever evolved, because—*by the very nature of male/female sexual reproduction,* coupled with *the very nature of evolution itself*—it couldn't have.

Ironically, *by the very nature of this book* I find myself repeating that basic thesis in various permutations over and over again, which is as frustrating to me as it may be annoying to you. In draft after draft, I've tried to eliminate unnecessary repetitions, yet in making each separate argument you are about to read, the particular point being made wouldn't be complete without my repeating one or more of the recurring fundamental problems. In each new context, it's crucial that we be reminded of the multiple impossibilities involved, given both the unique requirements for sexual reproduction and how evolution itself works (or, more to the point, *doesn't* work).

If use of the word *impossible* still sticks in your craw, I call your attention again to Darwin's own reference to the possible and the impossible: "If it could be demonstrated that any complex organ existed, which could *not possibly* have been formed by numerous, successive, slight modifications, my theory would absolutely break down."

The book you hold in your hands is doing nothing more than taking Darwin at his word, and arguing that there is indeed something that could *not possibly* have been formed by numerous, successive, slight modifications. That particular something, in various incarnations, is the appearance of reproductive sex when and where it would have been absolutely vital along the purported path of evolution. If the premise of this book is right, by Darwin's own test his theory absolutely breaks down, and with it the captivating, but scientifically-untenable Evolution Story.

INTRODUCTION

The great tragedy of Science: the slaying of a beautiful hypothesis by an ugly fact.
—Thomas Henry Huxley

In the beginning, SEX did not exist. Evolution theory teaches that the first organisms simply copied themselves. So normative, gendered sex as seen throughout Nature could not have begun without the appearance of the first-ever male and female organisms, mating in a never-before-seen way, and reproducing by a revolutionary method of reducing their chromosomes precisely in half then blending those halves together to produce one-of-a-kind offspring. How those first-ever sexually-reproducing organisms possibly could have evolved *before sexual reproduction existed* is quietly admitted by evolutionists to be the "Queen of Evolutionary Problems."

In its own struggle for survival in a world of competing ideas, the theory of evolution is proved unfit, not by its insightful observations of what actually occurs in Nature, but by a seemingly insignificant detail that could not possibly ever have occurred in Nature. Simply put, evolution obviously happens, but evolution cannot explain either the origin of sex or the exclusive pairing of unique, male/female sex in each of millions of species. For Darwin's Grand Theory of microbe-to-man evolution, those yawning gaps are an insoluble problem.

In this book, I'm not concerned about the so-called "missing link" between apes and humans. Nor am I primarily interested in the skeletal remains of "Lucy," the ever-intriguing early dinosaurs, the age of the universe, or the "six days" of Genesis, or Creation versus Evolution generally, or even Intelligent Design versus Evolution. Those books from every side have already been written. What we must understand from the outset is that the single issue presented in this book has more to do with logic and critical thinking than sophisticated science. If credible science could explain the apparent conundrum, there would be no problem to be thinking critically about.

Reiterating the scientifically obvious, I believe living beings evolve. Just walk with me through the Cotswold villages in England where I have lived much of my life in-between teaching stints (and am, even now, writing these words). Every time we enter through one of the many low doorways in the quaint shops and cottages in this green and pleasant land, we literally walk through evolutionary history. Despite the familiar warning signs saying "Mind Your Head," I can't tell you how many times I've painfully bumped into human evolution! Where have all those once-shorter Brits gone? And just how many breeds of dogs do I encounter on my walks along the lovely Cotswold paths, and how many varieties of roses grace my daily ramblings? I say again: living beings evolve, and in far more profound ways than simply the varying height of humans or fascinating breeds of dogs and roses.

So why question human evolution? Because when most folks today think of *human evolution*, they're not talking about a given population at times shorter and at other times taller, but about the Darwinian hypothesis that humans have evolved from lower primates which themselves evolved from the same common origin as fish, birds, and plants—all of which, over eons of time, evolved from some primitive, single-celled asexual prototype. This dual definition of "evolution" has worked great mischief in our often-heated conversations about the subject. If the thesis of this book is anywhere near correct, the first, more-limited usage of human evolution ("bounded evolution" *within* easily recognized classes of beings) wouldn't be disturbed in the least; but the second usage—the "unbounded evolution" so essential to Darwinism's bedrock assumption of common descent—could not survive.

Genuine Evolution Theory Versus the Hyped Evolution Story

That distinction could not be more important. This book is not challenging the legitimate science of evolution so useful in scientific research, healthcare, and technology, only the highly-romanticized, commercialized, politicized, and subsidized microbe-to-man Evolution Story. Without the indispensable sexual transitions required for that familiar story, there's simply no story to be told. What's more, the highly-touted Evolution Story of legend and lore contributes nothing of value whatsoever to scientific research or medical breakthroughs. To the contrary, any scientific methodology modeled after the elongated time-scales, hallmark randomness, and wild guesses of the hyped Evolution Story would be a disaster for practical scientific progress. Anyone willing to wait millions of years for Nature, acting randomly, to come up with a cure for cancer?

The Evolution Story of popular books and attention-grabbing headlines is not the *predictive* model of useful evolution theory, but the *pretentious* model of useless evolution fantasy. By unfortunate irony, the validity of legitimate evolution theory has caused many scientists working in the trenches of research and medicine to uncritically accept the bogus Evolution Story, repeating Darwin's own fundamental mistake: extrapolating wrongly from the clearly observable to the highly-speculative unobservable.

Question: Does it matter to current scientific research whether evolution theory wholly fails to explain the crucial sexual transition from amphibians to reptiles, or can't possibly offer a plausible explanation for the first-ever male/female pair of the praying mantis, complete with their bizarre cannibalistic sex? Why, then, does the scientific community cling so desperately to the Evolution Story when it serves no practical scientific purpose? Well, that's another story—a story often more about philosophy than science. In the pages ahead, by contrast, it's the flawed science of the fanciful Evolution Story that's of immediate interest.

In the end, it's not so much Darwin's origin of *species* that matters, but first and foremost the origin of *sex*, without which there could be no sexually-reproducing species of any kind.

What's Crucial in This Book

This book is tightly focused on two simple questions, the first being: *How could asexual (non-sexual) replication have evolved by natural selection into fully-gendered sexual reproduction?* And the book's thesis is equally simple: *If there is an unbridgeable gap between non-gendered asexual replication and male/female sexual reproduction, then that Achilles' heel is the fatal undoing of microbe-to-man evolution.*

Note that this is a HOW question, not a WHY question. It's easy to become distracted by all the buzz about sexual reproduction's counterintuitive advantages over asexual replication, as if that somehow explains the evolutionary mechanism that could have brought about all that beneficial sex in the first place. It's possible, of course, that knowing WHY could lead to knowing HOW, but you'd be surprised how often the WHY question is addressed with nary a nod to the HOW question. When that happens, we end up with a circular argument that assumes sex and then points to its supposed evolutionary advantages as if that explains its existence. Since sex itself would have to *exist* before it could be advantageous (or even be naturally *selected*), all such discussion is a pernicious shill in a high-stakes shell game.

For the limited purposes of this book, I'm going to join with the scientific community in its own typical starting-point: assuming the existence of asexual forms of life (replicating by a process of *mitosis*), however they actually came into being. Left in issue—the central issue of this book—is whether the scientific community is right in asserting that all sexually-reproducing life forms (which use a distinctively unique process known as *meiosis*) gradually evolved from asexually-reproducing life forms. In Chapters 4-6, we'll explore this pivotal issue in greater detail, but let's pave the way for that discussion with a number of preliminary considerations.

Among biologists, "sex" can sometimes mean nothing more than the exchange of genetic information, by whatever means. With that broadly-inclusive definition of sex, even the process of reproduction by such asexual organisms as bacteria and yeast is sometimes said to be "sexual reproduction." Along the way, we will consider a number of these intriguing "para-sexual" life cycles (all of which have insurmountable difficulties of their own regarding origins). But to prevent any misunderstanding, we need to carefully distinguish between the *purpose* of "sex" by any definition (the mingling of genes) and the *mechanisms* of sex (the physical means by which organisms mingle their genes). Since, either way, the end result is reproduction of some kind, one could easily conclude that the particular mechanisms themselves don't matter. Ah, but they matter a lot to true sexual reproduction, considering that without those particular mechanisms (whether it be compatible organs for mating or the unique reproductive process of *meiosis*) evolutionary sex wouldn't get out of the starting blocks, nor certainly ever reach the finish line.

If we can be frank for a moment, let's talk about the male penis. By evolution theory, no method of mingling genes would initially have involved a penis (much less a *male* penis, since at the beginning of evolution there would have been no males or females of any kind). So, the question is: How could any reproductive process which involves the male penis have evolved from any other reproductive process which functions without a penis? If there's no evolutionary mechanism to move from non-penile reproduction to penile reproduction, then none of the countless species which rely on penile reproduction ever could have evolved, calling a sudden halt to the notion of common descent. When it comes to the "Big-E" Evolution Story, the particular, specific mechanisms of sex matter hugely.

We could talk about any form of "gene mingling" that might be of interest, but in order to most sharply highlight evolution's multi-faceted mechanism problem, in this book we're tightly focusing on the most easily-recognizable and ubiquitous form of sexual reproduction. Call it "real sex." Real *male* and *female* sex. Real male and female *meiotic* sex, typically including the involvement of sperm and eggs.

While evolution's sex problem is equally problematic in other forms of gene mingling, never is it more clearly demonstrable than in this narrow focus. Make the case here, and it's sufficient to prove the point. More yet, it's sufficient alone to meet Darwin's own test for the undoing of his theory.

So, the question is: How did that specialized, unique, radically-different mode of reproduction ever come to be—especially considering the complication of germ cell differentiation into male and female?

Some will surely insist: "There are all sorts of creatures that reproduce using less than full-blown male/female reproduction, so who's to say that 'real male/female sex' could not have evolved in stages?" Yet, anyone familiar with those creatures will frankly concede that there are daunting problems *with each proposed stage*. More important for our immediate purpose, we're talking about a huge leap between the most advanced stage of single-cell "sex" one can imagine and full-blown male/female meiotic reproduction *in which it's not just a matter of exchanging genetic information, but a precise 50% reduction of chromosomes in both distinct genders, followed by a mind-blowing, intricate process of "crossing over."*[a]

When it comes to having sex, there's always a first time. This particular "first time" must be a prototype life form fully capable of *having sex*

[a] In case anyone should wonder, the many references hereafter to "*male/female* meiosis"—a phrase which normally would be redundant—is an attempt to carefully distinguish this particular, distinct form of sexually-reproducing meiosis from any form of asexual replication with which the word "meiosis" is associated. (Where the term "meiosis" is used alone and unqualified, it is shorthand for "male/female meiosis," unless the context indicates otherwise.)

and thereafter *sexually reproducing*. We're not talking here about human sexuality, complete with all the external and internal sexual apparatus we normally take for granted. We're talking about the ability to produce a *male* and a *female* (don't rush past that not-so-minor detail!) who *mate sexually* (don't take that for granted!) and then reproduce in a radically different (even counter-intuitive) way from any and every form of asexual replication that previously might have been in play. So, it's not just *one* "first time," but *three* "first times"—somewhere along the line *all at the same time*. And *all in the same place*. That's a huge ask!

The "Generation One" Requirement

With all the evolutionary talk about millions and billions of years, we tend to forget that time consists of an infinite series of "freeze-frame" moments along the length of time's continuum. If Darwin's Grand Theory (or even Neo-Darwinism) were true, there must have been a single, defining moment somewhere along the supposed evolutionary timeline when the unfolding of life (as supposed) moved from exclusively non-gendered asexual replication to male/female sexual reproduction for the very first time. In this book, I'm suggesting that we need to slow down evolution's whirring time machine to a single, uniquely-critical "freeze-frame" moment.

If you were directing the filming of evolutionary history and wanted to showcase a particularly-dramatic moment, you might insist on "stop-action, slow-motion," frame by frame by frame, with each frame being a super-tight close-up shot. After all, the old movie film was nothing more than individual snapshots connected together and run at such speed that it tricked our minds into thinking there was a "motion picture." The idea here is to reverse that process in order to make sure our minds aren't being tricked regarding something far more important than a matinee movie.

If the "unbounded evolution" hypothesis of progression from the simplest organism to human beings via natural selection is right, as a matter of actual historical fact there must have been a "freeze-frame" moment in which—*in a single "snapshot generation"*—two completely novel different kinds of replicators, one male and one female (having somehow evolved from one or more genderless, asexual precursors), must have been capable (in one go-or-no-go generation) of independently reducing the number of their respective chromosomes by precisely half. Then in that same "snapshot generation" both the male and the female must have joined in blending those chromosomes together to permit (yet again in one go-or-no-go generation) even the simplest form of sexual reproduction, ending up full-circle with offspring having the full number of chromosomes of each parent. If any essential part of that revolutionary process was missing, conceivably some other form of replication might have evolved, but not the ubiquitous process of male/female meiosis required for sexual reproduction as we know it.

Why must all of those contributing factors have come together in some single generation? (Surely, you can see it already....) Without a fully-developed (even simple prototype) *male* and a separate, fully-developed *female*, there can't possibly be *sexual* reproduction by any meiotic organism using a male/female reproductive process. *That goes double for the first-ever penile/vaginal sex.* And without sexual reproduction in a single generation, we can't possibly move on to the second, third, and fourth generations of sexually-reproducing life forms.

Would you be willing to bet that all these necessary occurrences could come together simultaneously in a single generation *if your life depended on it*? As it happens, your life *does* depend on it. If even one part of the process never happened, you wouldn't be here! Nor your parents, nor theirs. Nor the chickens, cows, or fish you eat. Nor the flowers or vegetables in your garden. Without 1) gender, 2) sexual interplay, and 3) sexual reproduction—all together in time and place in one generation from exclusively non-gendered, non-sexual duplication, it's "one, two, three, we all fall down."

At the crucial point of transition *somewhere in real time*, not even eons of incremental evolutionary gradualism could possibly be of any benefit. Merely consider the worthlessness of half a penis or half a vagina (meaning, not yet fully evolved or capable of functioning) and extrapolate that reality back to even the simplest of asexually-replicating life forms on their way to becoming sexual reproducers. *When it comes to the world's first genuine sexual experience, there's simply no time for preliminaries.*

Please don't hang up on me quite yet. I fully appreciate that no evolutionist of any stripe contends that such a single-generation transition event ever happened, nor does anyone suggest it could ever have happened. Virtually every evolutionist would agree that, if this one proposition were true, then, of course, "Big-E" Evolution would be unable to explain the emergence of gendered sex. But with all the force they can muster, they would insist that evolution simply doesn't work that way…and I wholeheartedly agree!

This may surprise you, but in the pages ahead (particularly in Chapter 4), I hope to convince you that evolutionists are absolutely right—that such a single-generation transition event could not possibly have happened because that's not how evolution works. Problem is, because of the way "little-e" evolution *does* work, "Big-E" Evolution *doesn't* work.

Evolutionists Have No Answers

It's not as if I've discovered a flaw in Darwinian evolution that no one else has ever thought of. In fact, those (relatively few) evolutionists who openly acknowledge the problem have given it the name, "The Queen of evolutionary problems." Yet, most others have pretty much ignored the problem. To pay it any serious attention is to risk the entire Darwinian house of cards crashing down. Every now and then, an evolutionist here or there will crack open the door ever so slightly, but then quickly shove "the Queen" back into the closet, either pretending nothing is amiss or thinking that surely *someday* someone will come up with a good explanation.

Even Richard Dawkins, the Darwin of our day and author of the best-selling classic, *The Selfish Gene*, readily admits that evolution doesn't have an answer to what might be called "the missing-sex-link problem." With commendable candor, Dawkins acknowledges the conundrum:

> The other assumption I have glossed over, that of the existence of sexual reproduction and crossing over, is more difficult to justify.... Why did sex, that bizarre perversion of straightforward replication, ever arise in the first place?... This is an extremely difficult question for the evolutionist to answer. Most serious attempts to answer it involve sophisticated mathematical reasoning. I am frankly going to evade it except to say one thing... [whereupon he reverts to his argument that genes were the original replicators, not individuals or groups].[1]

Dawkins then concludes with the facile circular argument that "if sexual, as opposed to non-sexual, reproduction benefits a gene for sexual reproduction, that is a sufficient explanation for the existence of sexual reproduction."[2] Does Dawkins not realize that before sex can be beneficial for a gene, the organism itself must first be sexual—begging the question of the hour: How did the very first genes associated with male/female sexual reproduction ever come into existence?

To his credit, Dawkins is forthright in admitting his circular reasoning, saying

> This comes perilously close to being a circular argument, since the existence of sexuality is a precondition for the whole chain of reasoning that leads to the gene being regarded as the unit of selection. I believe there are ways of escaping from the circularity, but this book is not the place to pursue the question.[3]

As far as I can tell, Dawkins has not subsequently shared with his readers those possible ways of resolving the problem. In *The Blind Watchmaker*, Dawkins once again punts, saying

> For reasons that I haven't the space to go into, the existence of sexual reproduction poses a big theoretical puzzle for Darwinians.[4]

In this book, we will further highlight the circular reasoning that allows evolutionists the luxury of evading evolution's fatal flaw. I cite Dawkins' important concessions here, first of all, to confirm that the origin of sex is indeed a problem for evolutionists; second, to show that even a superstar luminary like Richard Dawkins has no ready answer for the problem; and, finally, to lend a modicum of credibility to the central thesis of this book. This book didn't just fall off the turnip truck!

What We Can Know, and What We Can't

Dawkins isn't the only evolutionist to acknowledge that there is no readily identifiable answer to the evolutionary sex conundrum. In book after book and article after article, the bottom-line conclusion from those scientists who've seriously thought about the problem is always the same: "We don't know, and we know we don't know." Best we can say is, "Evolutionary sex is an enigma;" or "It's a mystery." Fair enough. Life is full of unexplained, and likely unexplainable, mysteries that we somehow manage to live with. Yet, that honest admission has the potential to work unseen mischief along two separate fronts.

For more-militant evolutionists, evolution's sex problem simply doesn't matter. Since, by evolutionist thinking, there's such overwhelming evidence that the Evolution Story is true, there's absolutely nothing that could possibly make it *not* true! "Maybe some day we'll have a break-through explanation, but it's not a theory killer if we never really know." After all, if the truly pivotal mystery of the origin of life from non-life is no big deal for evolution theory, what lesser mysteries possibly could overturn science's most sacrosanct dogma? (Well, actually, the mystery of sex.)

For evolutionists of all stripes, the lack of an explanation for something so crucial to evolution theory ought to raise red flags about the viability of that very theory. If we admit that we don't have a solid explanation for how sex evolved, how can science proclaim with its usual aura of infallibility that the Grand Theory—so dependent upon sexual reproduction—is the indisputable explanation for everything we observe in Nature, including (to go full circle) sexual reproduction itself? Honest confessions of ignorance about evolutionary sex are laudable until that ignorance becomes the weak link that breaks the entire chain of evolution.

And then there's this question (pursuant to the bold pitch in the opening Preface): Could it be that evolutionists have no explanation for the origin of sex because—given the very nature of male/female meiotic reproduction—an evolutionary explanation is simply not possible? If, despite your best efforts, you can't explain how pigs could possibly fly, maybe it's because it's *impossible* for pigs to fly. And equally impossible for evolution to provide a bridge between genderless asexual replication and fully-gendered sexual reproduction.

"Of course, it can!" I hear evolutionists saying. "There are any number of possible pathways." Then why the inevitable admission: "We don't know, and we know we don't know"? Maybe that means nothing more than "We don't *yet* know," or "We've got lots of good ideas, but don't know *for sure*." I understand. But if the door is to be left open for virtually any and every scientific possibility no matter how implausible, all I'm asking is a fair shake at your seriously considering one other possibility: the possibility of *impossibility*.

A Far More Widespread (and Even-More-Obvious) Sex Problem

While the necessary leap from asexual replication to sexual reproduction is the Grand Theory's most fundamental and insurmountable flaw, the problem of evolutionary sex might actually be more easily grasped when we consider the impossibility of natural selection evolving the first progenitors for millions of distinct species along the supposed line of evolution's common descent. Indeed, this is a second, stand-alone theory-buster wholly apart from the origin of sex itself.

The starting point couldn't be simpler. How do we know that we've got a distinct species? When it can't reproduce with any other organism, no matter how superficially similar. As wildly prolific as sex is throughout Nature, it is not species-promiscuous. In more than the obvious ways, sparrows can't reproduce with rhinos, or oak trees with apple trees.

Herein lies the second central question of this book, having to do with the Grand Theory's linchpin assumption of common-origin evolution (which assumes that all "higher" species evolved from "lower" species). *How could natural selection possibly have provided simultaneous, on-time delivery of both the male and female of each species, complete with their own unique, exclusive mating and reproductive processes, without which there could have been no second generation of that species...nor any next-higher species?*

When formulating his Grand Theory, Darwin overlooked (or ignored) this most obvious of all problems. There's not just one crucial gap at the origin of sexual reproduction itself, but in fact millions of gaps—one at each critical juncture where sexual reproduction is stubbornly species-unique, requiring a compatible *first pair* of fully-developed male and female forms of each species. Need that point even be argued? (Since apparently it does, you'll find that argument in Chapters 1-3.)

How very odd, then, (and how very telling) that in the vast collection of books and articles exploring evolution, there is virtually no mention whatsoever of this problem beyond an occasional reference to mystical, magical "coevolution," which is never further explained or explored. Can an entire scientific community be so oblivious to the obvious?

It's Not Rocket Science

When it comes to human reproduction, no scientist is baffled. If there were no males, human reproduction would be a non-starter. Same if there were no females. Talk about artificial insemination all you want, or even cloning these days, but none of that takes away from the obvious. Males don't produce eggs, females don't produce sperm, and—by force of nature—females without males don't produce babies. When it comes to your basic male/female reproductive sex, there's no deep, ponderous scientific mystery.

Now back up that human reproductive process to any "lower" species which also reproduces by male/female sex. Take robins, for example. Will there be offspring to perpetuate the species of robins if in the first generation there is no *male* robin? Or no *female* robin? Or if they are not compatible for mating? Or don't have the right internal processes of fertilization and reproduction? Surely, this isn't a complicated problem. Is there some scientific enigma beyond the obvious?

Now back up that same process to any still-further-removed precursor species that also reproduces by male/female sex. And then to the next, with all the same questions. If you're convinced that all "higher species" evolved from "lower species," get low. Get real low. Get down to the last (or *first*) single common denominator of all male/female reproducing species. At that point, as always, what all do we need? You're right. Exactly!

So, here's the big question: Is there even a remote possibility that natural selection—acting blindly—could provide simultaneous on-time delivery of both the very first male and the very first female, each fully fitted with the unique, compatible chromosomes and meiotic processes necessary to produce offspring capable of perpetuating that first species?

No scientific study on the origin of sex has come close to solving that problem, nor is anyone claiming such a prize-worthy discovery. As stated in the Preface, far from insisting that science immediately come up with a definitive explanation, the premise of this book is that—given the very nature of male/female sex—no study could possibly solve the problem.

Brash as it sounds, it's an *impossibility problem.* Natural selection can do some things, but natural selection can't do everything. *Of the things that natural selection absolutely could never have done, the most obvious and most crucial is to provide sexual reproduction when, where, and how it would have been required in order for the Grand Evolution Story to be true.*

Male/female (meiotic) reproductive sex isn't like anything else. It dances to a different tune. It has its own rules. Asking it to do what it can't possibly do is asking too much. So, to believe that someday science will come up with a credible evolutionary explanation defies all reason and logic. No matter how many studies you have, the results will be the same: pigs can't fly, rocks don't have sex, and the sex necessary for evolution cannot possibly have been produced by evolution.

Which is why, in the final section of the book (Chapters 11-15), we will address the attempt on the part of a growing number of folks to reconcile their religious beliefs with the entrenched Evolution Story in what is known as "evolutionary creation" or "theistic evolution." Based as it is on flawed science, that view is a futile, indeed unnecessary, attempt to reconcile the irreconcilable. Not even throwing God at the problem solves the problem.

Whether for believers or unbelievers, sex is not just the Holy Grail of evolution science, but the Wholly Fail of evolution theory.

PART ONE

Millions of Missing Sex Links

CHAPTER 1

Species and Sex: Catastrophes at the Intersections

[One's] imagination must fill up the very wide blanks.
—Charles Darwin (letter to Asa Gray)

Pointing to the millions of "missing sex links," in which each species requires its own unique, sexually-compatible male and female in Generation One, considering particularly the built-in constraints that govern outer boundaries and limitations of form and function.

Before we discuss evolution's most fundamental fatal flaw—the origin of sex—let's start with a closely-related flaw that's more familiar and far more easily understood. This flaw comes at the juncture where, by evolution theory, "lower" species would have evolved into other, "higher" species along a never-ending line of common descent. The problem is that natural selection could not possibly have provided simultaneous, on-time delivery of the first fully-compatible pair of male and female prototypes of any given species, much less for millions upon millions of sexually-reproducing species.

It's not so much a matter of complicated science as simple, common-sense logic. *One* won't do; gotta be *two*. And not just any two, but a unique set of male and female prototypes having sex and reproducing in a way unlike any other species on the face of the earth. In fact, it's that unique, tamper-proof sexual exclusivity that best defines a "species." Can you mate and reproduce sexually? Then you must be of the same species. Want to have sex and reproduce with another species? Sorry, wrong number!

And then there's that other glaring problem for evolution's famed gradualism. Given what it takes for one generation to produce the next generation (requiring everything in place all at once—compatible gender, unique mating, and one-of-a-kind reproduction), no intermediate forms possibly could advance the ball. No half-a-male of the new species; no half-a-female. No half-way novel mating equipment or partially-developed reproductive system. Unless you have everything—*the right kind of everything for each sexually-unique species*—you might as well have nothing. As you think about it, it's simple as pie: *One generation can't pass on any slight evolutionary advantages to a generation that can't yet be produced by slight advantages.*

Keeping it simple, humans obviously can't have sex with fish; birds can't do it with mice; and a rose can't possibly get it on with a rhinoceros. Among sexually-reproducing living forms, there are, incontrovertibly, fixed limits to mating and sexual reproduction. And according to all the available evidence, that has always been the case. Despite all the hype, colorful illustrations, and bold assertions, the fossil record virtually shouts that there are no intermediate forms whereby these sexual boundaries have ever been crossed. An indisputable fossil record, you say? Not for sex!

In *Eros and Evolution*, Richard Michod is candid about the degree of difficulty this problem posed for Darwin, and continues to pose:

> For Darwin, the question of the origin of species was the same as "Why are not all organic beings blended together in an inextricable chaos?" There is order in the living world; it is not an inextricable chaos. Species are, for the most part, distinct and recognizable; we give them names. Cultures

> without the formal study of modern biology name the same species as do trained biologists....
>
> Species membership is usually not a matter of degree but of discontinuity. No intermediate types of organisms typically exist with characteristics in between the characteristics typical of two good species—especially when these species are similar and found together. The gaps that separate different species are as basic to the biological species concept as are the characteristics shared by members of the same species.[5]

Evolutionist Mark Ridley also confirms the obvious—"that living organisms do occur in fairly discrete units, within which they can interbreed with almost equal efficiency, and outside which they can scarcely interbreed at all."[6] Indeed, Ridley joins with Stephen Jay Gould and Niles Eldredge in observing that, "fossil lineages generally appear suddenly in the fossil record, persist for a few million years, and then disappear abruptly without merging into any later lineages."[7]

Why should this be if all sexually-reproducing life forms gradually evolved from a common sexual progenitor? If evolutionary gradualism happened on a grand scale from fish to humans, at what point would that gradualism have abruptly stopped in millions of various life forms, whether high or low, such that there would be fixed sexual boundaries? How would each incrementally-higher species have come to exist, much less reproduce sexually to move evolution forward to higher and higher life forms? If, like Eldredge and Gould, you're a fan of "punctuated equilibrium" (saltational "leaps"), how do you explain the millions of separate sexual "leaps" that would have been required to end up with the unique, non-interbreeding sexual systems so ubiquitous throughout all of Nature?

Think of each new species as a "black spot intersection" where danger lurks for Darwinism as it travels the long and winding road from the primordial cell to humans. It's because fatal accidents frequently happen at "black spot intersections" that we see the warning signs as we approach. Evolution is all about accidents and non-directed occurrences, but if theoretically one "accident" might fortuitously result in evolutionary progress, an entire system utterly dependent on serial "accidents" can hardly explain the diversity required by millions of utterly fascinating mating habits and highly-complicated reproductive anatomies; nor, certainly, Nature's stubbornly-fixed boundaries that forbid interspecies reproduction. For the Grand Theory of evolution, these countless "black-spot intersections," like the ones on the highways, are not just problematic, but fatal.

Of Penises and Vaginas

The most obvious barrier to *having sex*, of course, is the no small matter of morphology (form). Different "species" (virtually self-defining on this basis alone) have distinct sexual anatomy that is incompatible with all but those of the same species. What could be more basic? If the male's sexual and reproductive apparatus isn't perfectly matched with the female's sexual and reproductive apparatus, then sex is a non-starter. (And that goes double for the male marsupial which has a forked penis used to fertilize eggs in the female's twin uteruses![8]) When it comes to sexual organs, functional compatibility isn't optional; it's required. In the words of the old rock song, "I hear you knocking, but you can't come in!"[9]

Alan Dixson gives us some technical insight into the compatibility issue:

> It is logical to suggest that penile and vaginal lengths have co-evolved in mammals. Selection has presumably favoured efficient placement of the ejaculate close to the cervical os, in order that spermatozoa may migrate (or be actively transported) into the uterous. The human vagina is approximately 5-6 in. (12.5-15cm) long...and is capable of dilation to accommodate the penis during intercourse.... Among the primates as a whole, there are only nine species

> for which both vaginal lengths and (erect) penile lengths are known, including *H. sapiens*. For this small sample there is a highly significant correlation between these two genital measurements."[10]

Surely, it doesn't take a genius to figure that, when it comes to sex, size really does matter! And not just for humans. "It is possible that the size of the female's swelling and the length of the male's penis have co-evolved in chimpanzees," Dixson says, mixing the obvious (compatibility) with the far from obvious (an almost magic-like cause behind that compatibility).[11]

Naturally, all this "coevolving" surely must have happened, because, after all...well...there it is! And where would we be without it! Says Dixson, "The reality is that in primates and other mammals the length of the erect penis and vaginal length tend to evolve in tandem."[12] Lucky us! Lucky first-ever penile/vaginal couple! It's certainly convenient (and dead necessary!) but just exactly *how* does all this "coevolving" happen? Any clues? Any suggestions? Any hint of hard, scientific evidence beyond pure speculation and sheer paradigm necessity?

Indeed, how did the first living beings whose descendants now mate by penile/vaginal sex ever "do it" for the very first time in order for the next generation to carry on "doing it"? This is the critical juncture in evolutionary development we previewed earlier: the dubious benefit of half a penis or half a vagina.

Please hear me clearly on this matter of sexual compatibility. When I refer (as I often will) to "half a penis, half a vagina," I'm not talking about *long* or *short* penises and vaginas. Individuals can reproduce successfully despite having highly imperfect matches. Men with egregiously short penises can sire children with women who have egregiously long vaginal canals (though with a possible loss in efficiency). So, the issue is not a short penis, but rather (assuming an evolutionary process) a *not-yet-fully-evolved* prototype *penis-to-be*. (Or not to be. That is the question!)

To fully appreciate this crucial point, let your mind trace back in supposed evolutionary history to a time when there's never ever been a penis or vagina in any species on the planet. Now think about the first-ever organism of any kind that would have used penile/vaginal sex for reproduction. Question: How did that first penile/vaginal pair come to have their historically-first-ever penis and never-before-seen prototype vagina? They're obviously not the descendants of any precursor organism having a penis or vagina, so these two completely separate and novel organs (both of them unique, but necessarily compatible in function) would have to evolve from some other primitive sexual organs, whatever they might have been.

Make that *gradually* by some random evolutionary process. The obvious problem is that if 1/100th of a never-before-evolved penis were somehow to evolve in a single generation of a precursor species, that "advantage" could not be passed along by succeeding generations so as to eventually become the first-ever fully-functioning, reproduction-ready penis of any size.[b]

But for purpose of argument, let's say that it could, and did. Now we've got ourselves the planet's first-ever penis. But without an incredibly improbable simultaneous, gradual evolution of the planet's first-ever fully-developed (and functionally compatible) vagina, the planet's first-ever penis would be all dressed up with no place to go! In terms of reproduction, it's useless. (And, of course, it would also be useless on its own without the entire supporting cast that make the anatomical bits function as they do. Sex is a production nightmare requiring a cast of thousands—anatomical, neurological, and psychological.[c])

Here's where the thesis of this book comes in. Unless and until we have *both* a reproductive-ready, first-ever penis *and* a reproductive-ready,

---

[b] At that point, natural selection—not being prescient—couldn't be less interested.

[c] If we had the luxury of time and space, the development of male and female internal and external sexual organs from the *same* embryonic structures would be worth a chapter in itself. The process is amazing!

first-ever vagina, sexual mating by penile/vaginal intercourse is a non-starter. And if that's a non-starter, then penile/vaginal reproduction is a non-starter. And without penile/vaginal reproduction, there will be no offspring having either penis or vagina to even remotely mate and procreate the next generation of penile/vaginal organisms. Without which, there could have been no process of evolution producing the countless species (whether "lower" or "higher") that reproduce by penile/vaginal sex. Which means we never get to...*us*!

This isn't the classic question of whether it's possible to gradually evolve something as complex as the human eye over multiple generations of different (*already-sexually-reproducing*) species. The first-time-ever ability to reproduce by penile/vaginal sex would not have had the luxury of multiple generations to evolve. More crucial yet, such a process would have required ridiculously remote chances of simultaneous evolution of *not one but two* distinct, never-before-seen, fully-developed sexual organs compatible for mating (along with all the other reproductive bells and whistles) as would be necessary to produce penile/vaginal offspring for succeeding generations. It simply couldn't have happened. Ever.

Does this not make eminent sense? Is not the logic absolutely watertight? Where is there any wiggle-room on this? No "study" is going to help us out here. Research over time isn't suddenly going to surprise us with an explanation. Imagine all you wish. Hypothesize to your heart's content. Speculate till the cows come home. *By the very nature of indispensably-conjoined penile/vaginal sex*, gradual evolution of the first-ever penile/vaginal sex is nothing short of *impossible*.

Why this prolonged, detailed focus on penises and vaginas? Because they are perhaps the most recognizable, understandable icons of the broader sweep of unique forms of reproductive sex found throughout millions of sexually-reproducing species from high to low, including those that don't mate using penises and vaginas. (Indeed, not even all penises are alike from species to species.) What's impossible in the evolution of penises and vaginas is equally impossible when it comes to millions of other (often bizarre) gender-distinctive, yet compatible, sexual features in novel prototypes of species from A to Z.

If the origin of sex itself is the quietly-acknowledged "Queen" of evolutionary problems, surely the "King" of evolutionary problems is coming up with the individual sexual features in millions of sexually-unique species. And do you not find this interesting? *Loads of studies have been done in an effort to explain the "Queen" of evolutionary problems, but one would be hard-pressed to find even a single study attempting to explain the "King" of evolutionary problems.* Is that possibly because anyone seriously thinking about it would realize the futility of such a study? That, *by its very nature*, unique conjoined sexual features in millions of distinct species defies evolutionary explanation? No study could possibly explain the impossible.

Sex Comes in Systems

And that's only for openers. We're reminded of the interdependent, systemic aspects of sex and sexual reproduction when Dixson hypothesizes about the interactive anatomical features that must also be considered in the supposed evolutionary development of sexual organs:

> ...the greater thickness of the human penis may be linked to changes which have occurred in the female pelvis and vagina during evolution: 'As the diameter of the bony pelvis increased over time to permit passage of an infant with a larger cranium, the size of the vaginal canal also became larger.' Co-evolution meanwhile selected for thickening of the penis, to facilitate 'a satisfactory fit' during coitus. As we shall see, this mechanical explanation is more credible than hypotheses concerning the effects of sexual selection via sperm competition upon human penile morphology.[13]

So, we're to understand that this "mechanical explanation" is more credible than the sperm-competition hypothesis? Actually, what's even more credible is that, mechanically-speaking, none of Dixson's conveniently-imagined adaptations possibly could have happened by evolutionary gradualism. If the mechanics aren't right for the *very first generation*, there will be no second, third, or fourth generations to ultimately get it right.

Having dismissed at the front door all possibility of the unthinkably miraculous, at the back door evolutionists happily supply their own version: "coevolution"!

As the Lord of Evolution, Darwin himself performed the miracle of coevolution, multiplying the loaves and fishes of gender through natural selection, not only to fit the various species to their places in Nature, but also, as Darwin put it, "to fit the two sexes of the same species to each other."[14]

Is it just me, or is the cavalierly-presented "tandem evolution" mantra just a bit too self-serving when you consider the unique, invariably-perfect, matching set of sexual and reproductive organs in the male and female forms of each species? It's certainly convenient being able to "imagine" the otherwise factually improbable (which imagination Darwin admitted was necessary "to fill up the very wide blanks"), but when did mere imagination without further evidence pass for hard science?

The Different Faces of "Coevolution"

Opening Carl Zimmer's book, *Evolution—The Triumph of an Idea*, I was excited to read in the Preface that he would be discussing "the coevolution of males and females." At long last, I thought, a knowledgeable evolutionist was going to explain how males and females of each species had evolved simultaneously. To my disappointment, Zimmer was referring only to his discussion of how flowers had co-evolved with insects and birds to perpetuate the sex necessary for flower reproduction. As everyone knows, flowers typically have both female organs (pistils) and male organs (anthers) which supply the pollen to fertilize the seeds. But, of course, there's usually the need to cross-pollinate from one flower to the next.

As Zimmer puts it, "Somehow the pollen from one plant must get to the eggs of another. And not just any other plant: it has to reach another member *of the same species*."[15] (What I've italicized is worth thinking about in and of itself.) Apart from species that use the breeze to spread their pollen, Nature must come up with some performance-perfect pollinators to do the job. Typically, that would be birds, bees, and insects.

If you want a really "Wow!" example, you could do worse than Zimmer's own featured Madagascar orchid named *Angraecum sesquipedale*, which has a petal shaped into an 11-inch (up to16-inch-deep) shaft, at the bottom of which is a reservoir of sweet nectar that is irresistible to one of Nature's believe-it-or-not wonders. Try picturing a moth with a tongue that coils up like a watch spring then suddenly stretches out to be (are you ready?) 16 inches long! As it reaches in for its prized nectar, it rubs its forehead against pollen grains that stick to its body until the next refreshing drink on an awaiting orchid ready to receive the pollen to fertilize its eggs.[16] Cool, huh?

Cooler yet, Darwin knew about this strange orchid and predicted that some insect eventually would be discovered that could perform this amazing feat. Sure enough, in 1903 scientists discovered the suspect hiding out on Madagascar, followed by other similar flowers and "buddied-up" insects in Brazil and South Africa. Given the orchid's seductive *modus operandi*, it was logical enough that Darwin might predict the existence of a pollinator precisely satisfying this extraordinary orchid's unusual criteria (otherwise the species would quickly die out). But how should we explain the incredible symbiosis? Are we to believe that there was gradual coevolution of the orchid and the insect independently over generations of adaptation caused by haphazard mutations in each species, with the happy coincidence that at the end of the process each species had precisely the right equipment to serve each other's needs—complete with that precisely-16-inch shaft in the orchid and that incredible 16-inch tongue on the tiny insect? Are you kidding!

We could stop there, dumbfounded and amazed, but if we focus separately—first on the orchid, then on the insect—we've got to figure out how each of them came into existence, regardless of those fascinating "16-inch features." By evolution theory, the very first prototype of that orchid must have evolved from some non-orchid plant that would not have produced the same type of "eggs" (much less all the microbiotics that go into making that particular type of "egg"). How, then, would that transition have happened consistent with the necessary "Generation One" rule whereby, in order to get to Generation Two of any new species, absolutely every novel, unique sexual feature would have to be in place simultaneously in Generation One?

And the same goes for the insect. Before it "evolved" that 16-inch tongue, the insect itself must have evolved from some different life form that reproduced in a sufficiently-different way as not to interbreed with other species. Somewhere back down the line, mating would have been impossible merely because of incompatible sexual equipment. So, what explains that giant leap in the crucial, indispensable First Generation? On the surface, coevolution sounds appealing. Reach down just the least bit further (say, sixteen inches...) and coevolution has a distinctly hollow ring.

Whether it's about the "coevolution" of plants and pollinators, predator and prey, penises and vaginas, or the very first male and female prototypes, the handy-dandy (if totally mystifying) solution of "tandem evolution" seems always to be at hand for evolutionists. And just how fortuitous is that? Merely to *say* it is to *prove* it!

If only the fail-safe mantra of "coevolution" made the Grand Theory easier to swallow. Instead, it merely points to the fact that, given the daunting dual-gender problem for each and every species, accepting Darwin's theory is *doubly difficult*...by a factor of millions!

The Devil is in the Details

The problem preventing interspecies sexual reproduction is not limited to incongruent sexual anatomy. Even within the internal reproductive systems, there is marked disparity in what might otherwise be considered "minor details." Consider, for instance, that there appear to be physiological constraints limiting the speed of sperm production within a given species.[17] How does evolution account for such a unique constraint? And how does evolution explain that chromosomes of different species always *look different*?[18] Which is simply another way of saying, as Ridley puts it: "Genes controlling interbreeding must change at speciation."[19]

Dig down even farther and the problems merely multiply. Without seeming to recognize the wider, negative implications for human evolution, the (militantly evolutionist) authors of *Biology for Dummies* speak volumes in a single paragraph:

> The most important thing you should know about mating is that members of different species can't successfully reproduce with each other. After all, the whole point of sexual reproduction is to create a new generation that contains the genetic information from the previous generations. Interspecies reproduction doesn't work because different species contain different numbers of chromosomes, and those chromosomes contain differing genes. For instance, humans carry 46 chromosomes in each cell, whereas chimpanzees have 48 per cell. If a human and a chimp were to mate, the cell divisions wouldn't be equal, and a theoretical offspring... probably wouldn't be able to survive.[20]

"Whatever happened to the chromosome[s] that disappeared from the great apes?" asks Sir John Maddox, who served for over 25 years as the editor of *Nature*, the prestigious journal published by the British Association for the Advancement of Science. Maddox has put his finger on a problem that must have occurred in countless separate species which, by Darwin's Grand Theory, purportedly evolved from species having different numbers of chromosomes.

It's worth noting that famed geneticist, Francis Collins, believes the difference in the chromosome number between the earlier chimpanzees (48) and later-evolving humans (46) easily resulted as "a consequence of two ancestral chromosomes having fused together to generate human chromosome 2."[21] Did this serendipitous fusion occur simultaneously in both a male chimp and a female chimp, permitting the first-ever generation of hominins, complete with novel sexual organs and reproductive systems unique to hominins-cum-humans...?

Whether chromosomes are being added or removed from one species to the next, such a change would be problematic for any transitional creature caught in limbo. Moreover, any gradual evolutionary process would entail destroying the progenitor creature along the way while never quite being sufficient for the incipient creature. There's also the problem of ensuring matching sets of chromosomes (ploidy), and the effect any chromosome differential would have on all the other chromosomes. Most important of all, how would evolution account for the exact same number of chromosomes in both the male and the female of each emerging species?

After all of his thoughtful hypothesizing about what *might* have happened, it's interesting to observe so distinguished a scientist as Sir John concluding that, "The understanding of speciation is not much deeper than in Darwin's time."[22]

The pesky "chromosomes problem" plagues every presumed transition from one species to another, and even sub-species having different chromosome counts from their supposed predecessors. (I use "species" here exactly like the *Dummies* authors use it.)

What, for example, explains why "the number of chromosomes per nucleus of each eukaryotic cell remains constant and characteristic *for the species* [my emphasis]"?[23] And, more importantly, why is it that "chromosomes...vary in numbers from species to species from two to more than a thousand...."?[24] If perhaps we could explain the doubling of chromosomes into a diploid state (say, as Lynn Margulis and Dorion Sagan have suggested in their hypothesis "that protists originally cannibalized other protists but did not digest them"),[25] what would explain why a given species might have more than a thousand chromosomes? Sounds like some serious cannibalizing going on!

And imagine what it would be like to step on the scales after all that cannibalizing: Whatever your "weight," it would always and inevitably be in exact multiples of doubles, from the first set of chromosomes (haploid) to diploid (two sets), to tetraploid (four sets) and so on. By what law of selection would the regularity of such "ploidy" (chromosomes in sets) have come into play? And what would explain why animals are diploid in their body cells, but haploid in their sex cells? It's no answer that "gametes are necessarily haploid because they fuse during fertilization,"[26] since that process itself only begs further explanation.

X and WHY Chromosomes

Speaking of sex cells, how does evolution explain the (necessarily simultaneous) origin of the X and Y sex chromosomes in humans and other animals? One hypothesis we're offered is that the X and Y chromosomes evolved from a pair of identical chromosomes when some ancestral creature developed a particular allele (genetic variable) which became the Y (male) chromosome while the other formerly-identical chromosome evolved as the X (female) chromosome.

Beyond the usual highly-speculative supposition involved, the problem is the same song, umpteenth verse. To be of any evolutionary benefit, all of this must have happened in a single generation, resulting in a *fully-developed* Y chromosome and a *fully-developed* X chromosome, each with all the male and female attributes respectively, each of which replicated with sufficient fidelity as to endure throughout every evolving X/Y-chromosome species, and on and on. Such an implausible scenario defies all rational (scientific) thought.

Not-So-Magic Mutations

Listening to many evolutionists, all it takes is a mutation here and a mutation there (or billions of them) and, *voilà*, life mysteriously moved from a single cell to us humans. The problem is that mutations are not just random, but are either neutral or more likely harmful. From that dismal prospect, the numbers game gets downright prohibitive. One or two random mutations are essentially worthless. In fact, Michael Pitman reminds us that "for evolution to occur through mutation, countless sequential good mutations would be required; at each step, all would have to cooperate harmoniously and each mutation would have to be selected for. This simply could not happen."[27]

In his (evolution-assuming) work on philosophical psychology, *The Ghost in the Machine*, Arthur Koestler put it rather more strongly, saying,

> Each mutation occurring alone would be wiped out before it could be combined with the others. They are all interdependent. The doctrine that their coming together was due to a series of blind coincidences is an affront not only to common sense but to the basic principle of scientific explanation.[28]

Evolutionist Mark Ridley agrees:

> Natural selection, however, can only change one part at a time, for the following reason. The chance of a single advantageous mutation is low enough; but the chance of simultaneous advantageous mutations, in the different parts of a complex organ, must be lower still. If the chance of one correct mutation is (say) less than one in a million, then the chance of ten correct ones would be impossibly small. For ten changes, the chance will be one in 10^{-60}, which is far too low a probability for the event ever to have taken place.... This being so, if there are complex organs that could only have arisen by exact changes in many different parts at the same time then they probably could not have evolved by natural selection.[29]

With all the talk about how mutations are the lifeblood of evolution, here are some caveats to consider: 1) For a mutation to be passed on to future generations, the mutation must occur in the germ cell, not just in the somatic cells.[d] 2) Given environmental differences, mutations are as likely to be selected *against* as to be selected. 3) There may be different probabilities of success, depending on whether mutations relate to substitutions, additions, or deletions. 4) Since mutations occur in organisms already well-adapted to their particular environment, the most likely outcome of any mutation is not positive but negative. 5) Mutations cannot add any new genetic information to an existing genome.[30]

All of which is to say that, while mutations obviously do occur and do effect change, mutations are not exactly what they're cracked up to be as any kind of "silver bullet" explanation for evolutionary progression from one sexually-exclusive species to another.

---

[d] This is much easier said than done. The process of meiosis itself almost ensures that mutations are eradicated and never passed on. Think of the number of sperm produced at any one time. Only the fully fit will ever get a chance to get to the egg.

One doesn't have to abandon logic and common sense to be a scientist or even an evolutionist. In fact, illogical hypotheses and dodgy explanations ought to send up red flags for scientists more than anyone else. Especially is that true when mutational explanations are based largely on lab experiments, such as the work of Morgan, Goldschmidt, Muller and others bombarding generations of fruit flies (*Drosophila*) with radiation, chemicals, light and darkness, heat and cold, to bring about mutations. Instead of providing proof of evolution, these experiments have confirmed what has always been known—that mutations are either insignificant or deleterious.

The rare mutations which survive outside laboratory conditions are mostly sterile or tend to revert to archetypical form rather than to evolve. In terms of natural selection, mutations are conservative, not progressive.[31] No matter how many mutations were artificially induced in the lab, fruit flies remained fruit flies. A little weird, maybe, but still fruit flies! Beware of mutation arguments glibly bandied about. *Mutations may sound sexy, but they have little to do with a sex-saturated world.*

As it turns out, *sex itself* may have far more to do with variation and speciation than rogue mutations. It's mostly down to the mixing and matching of genes. As the Ukrainian-born Theodosius Dobzhansky noted, even if two different species happened to mate, they rarely would produce hybrids. Why? Drawing from his arduous fruit-fly experiments, Dobzhansky demonstrated that the genes of one species simply clash with genes from other species.[32] That "clashing" is built in. It's "fixed," if you will. Nothing which overly-hyped mutations can possibly overcome. As Dobzhansky's findings highlight, the much-touted role of mutations in evolution is, like rumors of Mark Twain's death, greatly exaggerated.

Sex with Pin Numbers!

Just when you think it can't get any more complicated, the intricacies of sexual reproduction are seen to be all the more amazing. As the *Dummies* authors explain, a species may have its own "pin number" which it safeguards with its very life:

> Eggs are actually surrounded by a layer of proteins on top of the plasma membrane that contains receptor molecules made solely for receiving sperm of the same species. In human eggs, the zone that prevents fertilization by a different species is called the *zona pellucida*. Only human sperm can crack the code to get into the egg.[33]

That's mind-blowingly amazing. Let natural selection explain sexual "pin numbers"!

And did both the code and the receptor molecules "co-evolve" gradually over thousands or millions of generations? How would that have worked with the first female of some novel, distinct species gradually developing coded receptor molecules while a separately-evolving male of that species would gradually evolve so as to have the perfect code to unlock her secrets? Miracles may not be logical, but they are nothing near as *illogical.*

Along the same line, by what evolutionary logic did it come to pass that some electric fish send out electrical impulses *identifiable solely to members of the same species*?[34] The same goes for flowers, in which different types of stigma and pollen ensure that the female receives only the male of her species?[35] How could any of those "security measures" have developed in Generation One from the closest predecessor one might imagine? Indeed, the fact that species as different as fish and flowers should have analogous reproductive "security codes" preventing interspecies sex puts the probabilities of random selection somewhere beyond the moon. And multiply that by millions of password-protected species!

As you will hear me say repeatedly, description is not the same as explanation (as if merely describing something automatically explains it).[e] In this particular case, however, to describe in detail the intricacies of "security codes" and dedicated electrical frequencies within separate species is to confirm the very reason why there's a missing sex link between each and every distinct life form. The DNA for one security code can't pass on the DNA necessary for a completely different security code. This distinct, sexually-unique coding for each species is the reason why one life form didn't spawn any "higher" life form, whether in a single generation or over billions of generations—the same reason why common descent never happened, and the same reason why, at the end of the day, Darwin's Grand Theory simply doesn't work.

As Mendel demonstrated with his famous peas, almost endless genetic variations are possible with all those countless combinations of pea genes—far more than the combinations possible in, say, a deck of cards. But here's the kicker. While the possible combinations of cards within the deck may result in seemingly infinite permutations, *the deck itself doesn't change!* If in the deck of one species there are no genes that would be required for another fully-functioning species, no amount of shuffling of the deck is going to produce some other species. You can't pass on to others what you yourself don't have! Discrete species have discrete genes for discrete form and function. You can shuffle genetic information all day long, but the available pool of genetic information itself doesn't change, definitively ruling out the rise of genetically-novel species required of "unbounded evolution."

---

[e] Interestingly, this phrase is virtually the same as that used by Darwin's biographer, Peter Brent, speaking of Darwin's critique of Lamarck's "insubstantial vapour of theories that *described but did not explain*...." By contrast, says Brent (p. 305), Darwin's approach was to painstakingly test his theories against the realities of the natural world—though Brent was frank to say that sometimes "the line between testing the theory against the facts and finding facts to support the theory was hard to define and, even where defined, proved permeable."

A Final Warning About Those Black-spot Intersections

If you have a passion for further exploring the evolution of life within any given species (say, within the fascinating varieties of fish, or birds, or dogs, or spiders, or plants in the garden, or even observable changes within the human race), go for it! There's broad scope for wide-ranging research, creative hypotheses, and untold mysteries of Nature yet to be uncovered. It is crystal clear that evolution happens! And natural selection also happens. But both must be understood within the limitations of "bounded evolution"—especially the limitations of sexual reproduction we are highlighting in this volume.

What *doesn't* happen, and has *never* happened, is for sexual reproduction in one sexually-distinct species to evolve into a different form of sexual reproduction in another species. "Fish sex" can't evolve into "plant sex;" "bird sex" can't evolve into "horse sex;" and "dog sex" can't evolve into "cat sex" (even by the most indirect, "back-down-the-line" precursor links imaginable). Most importantly, "human sex"—even if there are similarities—is not simply aping some lower "primate sex." Nor were any the above, at any point in any way, derived transitionally from any common progenitor over however many billions of years anyone might wish to imagine.

In the first of his seminal *Transmutation Notebooks* leading up to *The Origin*, Darwin asked the pivotal question that we must ask ourselves: "Each species changes. Does it progress?"[36] To answer that question in the affirmative, as Darwin did, requires countless jumps from one internally-mutable yet "fixed" species to another internally-mutable yet "fixed" species—a phenomenon Darwin himself never observed but hoped against hope would be confirmed eventually by a more complete fossil record. To this day (except in the minds of evolutionists who fill in the missing gaps with their own imaginings) the fossil record continues to confirm the fixity of major species and phyla.

On the question of *progress* versus the *fixity of species*, merely consider Stephen Jay Gould's observation of the fossil record following on the heels of what he describes as 100 million years of life, from Ediacara, to Tommotian, to Burgess. "Since then," says Gould, "more than 500 million years of wonderful stories, triumphs and tragedies, but not a single new phylum, or basic anatomical design, added to the Burgess complement [only twists and turns upon established designs]."[37]

What do you think? Is that evidence of *progress* from lower to higher species, or is that the fundamental *stability* of recognizable classifications? And if someone protests that progress could be demonstrated if only given enough time, wouldn't you think that 500 million years is a fair test of that proposition? Yet from no less credible a witness than Stephen Gould comes testimony that, even given 500 million years, there are still no crucial bridges across the clearly-evident gaps among established phyla.

Have you ever considered the interesting parallel between gaps in the fossil record and gaps in scientific knowledge alluded to by Darwin himself at the end of his voyage on the Beagle? Lamenting that the necessary speed of travel precluded more exploration, Darwin confessed: "Hence arises, as I have found to my cost, a constant tendency to fill up the wide gaps of knowledge by inaccurate & superficial hypotheses."[38] Is there a gap in the fossil record? Somehow, some way, we've got to fill it! Is there a gap in a theory? Somehow, some way, we've got to fill it! Is there a gap between species? Somehow, some way—we can't help ourselves—we've got to fill it! Even, if necessary, with inaccurate and superficial hypotheses.

Is there evolutionary change? Yes. But that's *change*, not *progress*. Change, not ascending or descending species. Change, not a branching biological tree from some common ancestor. Change, not microbe-to-man evolution.

One should use extreme caution when approaching black-spot intersections. Fatal flaws in grand theories have been known to emerge from there.

Recap

- Each species is sexually unique, being defined first and foremost by its inability to mate and reproduce with any other species.
- To move from one sexually-unique species to another sexually-unique species, everything required for novel sexual reproduction would have to be in place in Generation One of each succeeding new species, especially any "higher" species along evolution's purported path.
- The usefulness of partially evolved mating and reproductive organs is ruled out by the fact that no generation can pass on any slight evolutionary advantages to a generation that can't yet be produced by slight advantages.
- Natural selection could not possibly have provided simultaneous, on-time delivery of the first fully-compatible pair of male and female prototypes of any given species, much less for millions of sexually-reproducing species.
- Given that inability, evolution's bedrock premise of common descent could never have happened, negating altogether the popular microbe-to-man Evolution Story.

Question: Would you still believe the popular Evolution Story if the millions of sexually-unique organisms in Nature could never have come into existence by a gradual process of evolution from precursor organisms having their own unique mating and reproductive mechanisms?

CHAPTER 2

Sex-Endangered Species

Extinction is the rule. Survival is the exception.
—Carl Sagan

Delving deeper into the problems surrounding speciation; examining built-in constraints; distinguishing between adaptation and speciation; and challenging explanations offered by evolutionists about the causes of speciation.

Do you not find all the chatter about endangered species especially interesting in light of evolution's linchpin doctrine of "survival of the fittest"? Doesn't evolution *assume* the importance of extinction along the path of evolutionary progress? The cuddlier the species, of course, the more we want to ensure its survival. But evolution can be brutal. It doesn't do cuddly. And Nature itself doesn't seem to be all that bothered about the loss of any given species. One can almost see ancient eco-warriors protesting the imminent extinction of dinosaurs, yet somehow we're managing to get along pretty well without them.

If ever there were a serious threat to species that ought to be protested, it is the cuddly notion that evolution's natural selection produced the millions of species that have existed over time, whether now extinct or still extant. In response, of course, evolutionists invariably ask, "Can we not, even now, see the evolution of species (and extinction) taking place right before our very eyes?" Sex isn't static or fixed, we're told.

Sexual mating and sexual reproduction are constantly in flux. And if it's happening *now*, why not from the very beginning?

As a case study of speciation, perhaps you've heard of the two species of European gulls—the *herring gull* and the *lesser black backed gull.* In Europe, the two species neither interbreed nor resemble each other. But if you trace the *lesser black gull* eastward, you will see a growing similarity with the *herring gull*; and if you trace the *herring gull westward*, you will see a growing similarity with the *lesser black gull.* So, what are the implications for interspecies sex? At the terminal extremes, the two forms of gulls are distinct, non-interbreeding species;[f] but in between, the various races of gulls do interbreed with adjacent races.[39]

Without question, then, speciation in Nature along a horizontal plane can be seen to happen. Yet just how far does this example move us along the supposed vertical chain of evolution from one species to another? Obviously not to a higher species, and certainly not to a radically different species. We're still talking gulls here, complete with intact, fully-developed *male* gulls and *female gulls*, and all the ancillary anatomical and physiological support systems characteristic of gulls. Along the plane of horizontal evolution, just as boys will be boys, gulls will be gulls!

As much could also be said for the pigeons with which Darwin was so fascinated, the varieties of which, at the hands of pigeon breeders, could change radically enough to look like entirely new species. "If you came across a fantail and a Jacobin in the wild," Carl Zimmer observes, "you might think they were different species, yet, strangely, they could still mate and produce fledglings."[40] Which is to say, they remain the same *general* "species," not some transitional, intermediate form on its way to becoming a duck or a chicken.

---

[f] This may have more to do with plumage than anything else. If there was herring gull IVF, then fertile offspring would be the result. They do not breed because they are no longer attracted to each other. They are dancing to different music.

Much the same goes for the fascinating fruit flies of Hawaii (and the hundreds of descendant species deriving from snails, moths, beetles, and wasps that are unique to Hawaii). Thanks to the oversized chromosomes in organs like the salivary glands, it's been possible to trace the order of genes in various *Drosophila* species on the different Hawaiian Islands so as to work out the entire evolutionary history of most Hawaiian species.[41]

There's no denying that on the level of "bounded evolution," new "species" of the same "kind" do evolve. But rather than documenting "unbounded evolution," the interbreeding limitations of "bounded evolution" merely highlight that "new species" (as in the case of gulls) are simply variations on theme, not variations which produce radically-different *novel* species. Whatever "new species" they might be, fruit flies are still fruit flies; snails are still snails; moths, still moths; beetles, still beetles; and wasps, still wasps. And that goes as well for Darwin's famed Galapagos finches. Longer or shorter beaks, perhaps, but still finches.

Bolstering that point, Gerald Schroeder points us to "The magnificent Natural History Museum in London [which] devotes an entire wing to demonstrating the fact of evolution. They show how pink daisies evolve into blue daisies, how gray moths change into black moths.... Impressive, until you walk out and reflect upon what they were able to document. Daisies remained daisies, moths remained moths, and cichlid fish remained cichlid fish.... The museum's staff did not demonstrate a single unequivocal case in which life underwent a major gradual morphological change."[42]

In any order you wish to start and finish, try charting evolution from fruit flies to snails, or snails to fruit flies; or moths to beetles, or beetles to wasps. No matter how hard you try, you can't get there from here! And you'll certainly not get the traditional (always-graphically-illustrated) evolutionary chain (or "tree") from fish to amphibian to reptile to mammal to man. As Michael Denton observes, no major divisions in

Nature could have been crossed by the same simple sorts of processes seen in various examples of evolution within species.[43]

This, in a nutshell, is the answer to the debate-portfolio question put by Mark Ridley:

> There is a continuous gradation from individual variation, through geographical variation, through such levels as the subspecies, through the species level, and up through the Linnaean hierarchy of genus, family, order, class, phylum.... The degrees of difference are a continuum. If any point on the continuum is chosen as the limit of evolution, a paradox immediately arises. If evolution can produce all the changes up to that point, why can it not produce the tiny change from one side of the point to the other?[44]

Did you catch the two false premises in Ridley's question? First, the circular reasoning that *assumes* continuous gradation, moving from lower to higher species along a continuum. That's the very question at issue! Second, is the highly-contestable assumption that evolution can produce all the necessary changes up to a given point. Given the "Generation One" problem, that simply isn't true. So, the answer to Ridley's question is that evolution can produce neither one side of the magic point nor the other. Variation within species is most definitely a difference in degree, but incompatible differences in sexual equipment and unique reproductive capabilities are differences in kind. Vary a species not just by degree but by sexually-unique kind, and it can kiss goodbye having sex with its closest neighbor.

The Problem of Constraints

Perhaps this seemingly-built-in incompatibility explains a mystery which Ridley himself confesses is an unresolved "problem" with evolution theory—namely, the *constraints* that prevent there being, say, legged snakes or insects with backbones. "Constraints do exist," Ridley acknowledges, but exactly how they work, he doesn't profess to know, other than guessing that the problem is at the level of mechanism.[45]

If such constraints *do* exist, doesn't that take the wind out of the sails of Ridley's previous argument that "if evolution can produce all the changes up to a point, why can it not produce the tiny change from one side of the point to the other?" Could not the answer be the very constraints that so bewildered Ridley—constraints at the level of mechanism which, like a rubber band stretched by degrees, can only stretch so far without finally breaking?

In *The Origin*, Darwin responded to critics who maintained that the amount of variation in artificial breeding has certain outer limits. "No doubt, as Mr. Wallace has remarked with much truth, a limit will be at last reached. For instance, there must be a limit to the fleetness of any terrestrial animal, as this will be determined by the friction to be overcome, the weight of body to be carried, and the power of contraction in the muscular fibres."[46] If you extend Darwin's logic to the complex intricacies of sexual reproduction and the myriad of indispensable supporting parts of an organism, the natural constraints against unlimited variation become even more apparent.

Relating to speciation in flowering plants, Darwin himself gives a good example of why even variations have built-in limits [with my emphasis in the following]: "It must not be supposed," he said, "that bees would thus produce a multitude of hybrids between *distinct species*; for if a plant's own pollen and that from *another species* are placed on the same stigma, the former is so prepotent that it invariably and completely destroys..."[47] (*Distinct species?* What makes them *distinct*? Having tried so hard to downplay any notion of distinct species, not even Darwin himself could fully ignore factual reality.)

One good way to think about biological constraints is to take a walk and look at the trees you see along the way. Why don't trees grow taller than they do? Why aren't the tallest Redwoods even taller? Invoking the limitations of physics merely begs the question of where *those* limitations came from. Looking closer, what prevents the squirrels and the birds in the trees from being twice as large as they are? The list of constraints is endless: Why are there usually only *two genders* throughout the plant and animal kingdom? Why do we consistently find only diploids and haploids in the process of meiosis? Why do we consistently find exact-copy duplication in the process of mitosis? Just how many crucial constraints throughout Nature need be cited?

Darwin himself begrudgingly acknowledged at least an evolved "fixity" of sorts when trying to explain why major organs (presumably including sexual organs) reach a point where they no longer evolve. "When a species with any extraordinarily-developed organ has become the parent of many modified descendants...in this case natural selection has succeeded in giving a fixed character to the organ."[48] Even if that convenient explanation were anywhere near correct about other organs, it wouldn't begin to explain the evolution of sexual organs which are simply too fixed in each species to propagate descendants having unique, sexually-exclusive properties.

If you truly want a constraint to ponder, consider what fixed principle of evolution would have dictated that all living creatures eventually should die? What scientific necessity explains the universal phenomenon of *death*?[g] Death is the *ultimate constraint*, making necessary the sexual reproduction which keeps life ticking and in harmonious balance. Ironic, isn't it? Natural selection needs death in order to function, but it can't explain death, only extinction. *Like Stephen Hawking's pretentious "theory of everything," evolution promises to explain everything from start to finish, but doesn't begin to explain the most important things.*

---

[g] "Organisms are programmed to die," says Richard Michod (*Eros and Evolution*, p. 59). Why? Because of sex, which "hastens the journey to the grave for all of us" (p. 52). How then explain the death of asexual organisms? As Michod acknowledges, "All organisms must die, of course." But if natural selection is about *survival*, does the universal constraint of death make evolutionary sense?

Given these biological constraints of the highest order, Darwin naturally had to struggle with the obvious "next question": Where did those constraints come from? Why did the various types (species) in Nature appear to be fixed?

Fixed "Tribes" Even If Mutable

Toying with the notion of stabilizing "monads" (which Gottfried Leibniz had postulated in his *Monadology*, 150 years prior), Darwin believed for a time that these mysterious monads were the ultimate units regulating the form (and even longevity) of any given species.[49] But it will surprise many to learn that, when writing the first of his Transmutation Notebooks, Darwin eventually rejected monads and actually reverted to divine creation, saying (with my emphasis): "*The Creator has made tribes of animals*, adapted preeminently for each element, but it seems law that *such tribes, as far as compatible with such structure, are in minor degree adapted* for other elements."[50]

Wow, the Creator made *tribes of animals*? That's some concession coming from Darwin! Still today, Darwin's "common ancestor thesis" keeps running headlong into his initial and incontrovertible observation: the stability of "tribes of animals, adapted preeminently for each element," necessarily including their distinct sexual organs, exclusive reproductive systems, and unique mating habits.

This observable stability in reproductive systems also explains why, when it comes to the tricky business of classifying organisms, the distinctive organs of reproduction are more reliable than any other parts of the body or such extraneous matters as creature habits and food. As noted in *The Origin*, even Darwin recognized the distinctiveness of species based on reproductive features:

> Owen, in speaking of the dugong, says, "The generative organs, being those which are most remotely related to the habits and food of an animal, I have always regarded as

> affording very clear indications of its true affinities. We are least likely in the modification of these organs to mistake a merely adaptive for an essential character." With plants how remarkable it is that the organs of vegetation, on which their nutrition and life depend, are of little significance; whereas the organs of reproduction, with their product the seed and embryo, are of paramount importance![51]

What did Darwin say was the most reliable key to identifying and classifying species? Sex! If there are different sexual organs, then we've got ourselves a different species. If there's a distinct means of sexual reproduction, then what we have is a distinct species.

When it comes to speciation, Darwin was certainly right about one thing: the organs of reproduction are of paramount importance.

Hox **Genes, or Hoax Genes?**

One can almost hear Carl Zimmer saying, "Hold on, what about the Hox genes that are found in animals as diverse as frogs, mice, and humans, doing the exact same job in all of those animals—namely, shaping the head-to-tail part of each developing body?" If "a defective *Hox* gene in a fruit fly is replaced with the corresponding *Hox* gene from a mouse...the fly will still grow its proper body parts."[52] Calling such genes "master-control genes," or "the common genetic tool kit," Zimmer as much suggests that the same "card" (the *Hox* gene) is found within all those various discrete "decks" (which, Zimmer posits, surely must have evolved in the various animals' common ancestor).[53]

How would you answer Zimmer's objection? Exactly! If there were a master-control gene for sexual or reproductive organs, how are we to explain such wide variety of form and function? Master-control genes may be "able to use the same body-building instructions to build very different kinds of animal bodies," but no commonly-evolved master-control gene could possibly account for the millions of sexually-reproducing life forms so diverse that they can't even remotely interbreed. (It's a longer story, but *Hox* genes alone don't actually design bodies, appearing on the scene, as they do, after the body plan is established. And since mutations in *Hox* genes don't have all the genetic information necessary for body-building, simply mutating them won't buy you a bus ticket to the next stop on the upward road of evolution.)

Directly pertinent to this book, sex (particularly at the microbiotic level) isn't as simple as head-to-tail body development along a certain axis. If a sex gene in a fruit fly is replaced with a corresponding sex gene from a mouse, you can safely bet that the fly won't produce a mouse, or even a "flouse"!

That's the fundamental fallacy of Zimmer's imaginative extension of the *Hox*-gene theory whereby he supposes that the 13 *Hox* genes in lancelets ("our closest living invertebrate relative") somehow multiplied four-fold into the number of *Hox* genes in vertebrates. "With some tinkering to their genetic tool kit, vertebrates were able to grow noses, eyes, skeletons, and powerful swallowing muscles." [Who's doing the tinkering? Nature?] Indeed, "thanks to their genetic revolution, the early vertebrates eventually gave rise to sharks, anacondas, humans, and whales."[54]

Anyone skimming Zimmer's mesmerizing explanation might easily miss the missing sex links. Look closely at what Zimmer says: "... other copies of *Hox* genes evolved until they were able to help shape the vertebrate embryo in new ways." *What vertebrate? What vertebrate embryo?* Mention the word *embryo*, and we're talking vertebrate sex here—a radically different form of sex and a unique type of embryo never before seen on the planet! Forget the noses, eyes, and skeleton. Whatever it might have been (a lamprey-like creature?), the very first vertebrate—make that the *two* first vertebrates (one male and one female)—had to emerge from the lancelet, fully formed and capable of mating and reproducing as those first vertebrates would have mated and reproduced.

There would be no time for *Hox* genes themselves to evolve, and—even if we supposed they did—it's not the *Hox genes* that really matter, only the progenitor lancelet's *sex genes.* No vertebrate sex genes in the invertebrate lancelet, no vertebrates. No vertebrate, no vertebrate embryo. No vertebrate embryo, no next generation of *Hox* genes to evolve all those vertebrate noses, eyes, and skeletons. And most importantly—if no vertebrate noses, eyes, and skeletons in the first vertebrates, then no ensuing sharks, anacondas, humans, or whales.

Hox genes, or *hoax genes*? It only takes the slightest of mutations to get from one word to the next.

Does "Speciation" Promise More Than It Delivers?

If you're looking for evidence of the origin of species, it's instructive to read Zimmer and Emlen's popular text on evolution, particularly section 13.3, "The Origin of Isolating Barriers: How New Species Form." After discussing geographic barriers and reproductive barriers, the authors acknowledge the obvious, saying, "The many different isolating barriers in nature help keep sexually reproducing species distinct. But these barriers had to come from somewhere."[55] Indeed they did, but the question is, *From where*? Or perhaps *How*? The authors' highlighted answer is geographic isolation:

> New species evolve when populations become geographically separated from each other.... The divided populations continue to evolve: new mutations arise in each one, some of which become fixed through genetic drift. Each population accumulates its own unique set of mutations, and the longer they are divided, the more of these mutations they accumulate.... [O]nce they have begun to evolve independently, reproductive barriers to gene flow can begin to accumulate.

But if populations that have been geographically divided come back into contact again, there is renewed opportunity to interbreed. "If the barriers are weak, the population will interbreed easily *with the rest of its species* [emphasis mine]." You mean *the same species*? But weren't "new species" developed when the populations were divided? Not only does this alert us to the loose use of the term "species," but more importantly it highlights the nature of what evolutionists call a "new species" evolving in the process of speciation. If the "new species" can mate with the "old species" at their high-school reunion, what obviously *hasn't* changed? Their compatible sexual equipment, right? And isn't it equally clear what else hasn't changed? The basic creature in question, whether a particular "species" of flies, shrimp, butterflies, or beetles.

Not a single "new species" presented by Zimmer and Emlen in the discussion of "The Origin of Species" (their Chapter 13) is anything other than a spin-off variety of the mother-ship creature. If perhaps there are pre-zygotic or post-zygotic barriers to reproduction caused by geographical isolation, the apparatus for mating hasn't radically changed, nor has the basic archetypical form and function of the given creature. This process of "speciation" is light years away from any process of speciation capable of producing millions of species that have radically-different mating and reproductive systems within radically-different living beings, as is assumed by Darwin's Grand Theory.

Hybrids Cars May Go a Long Way, But Not Hybrid Sex

Once you head down that unbounded path, speciation starts getting ugly. From Zimmer and Emlen's own text, we learn that "embryos with two species for parents may fail to develop." And "Even if two species have all of the same genes, they may have alleles of those genes that cannot function together." Then there are the cases of hybrids, which "sometimes develop successfully, but in some cases they are born with deformities that leave them in poor health." Or (working against the upward evolution of species) they are most likely sterile. "Hybrid male

Drosophila flies, for example, produce defective sperm. Mules, which are the hybrid offspring of horses and donkeys, are almost never able to breed with each other, nor can they breed with horses or donkeys... their sterility means that horses and donkeys remain distinct species."[56]

But wait, says Mark Ridley, "If the numbers of their hereditary structures called chromosomes can be caused to double, the hybrids can reproduce." Note the operative words, "can be caused." As Ridley insists: "New species can be artificially made from the old ones." *Artificially*... as in *purposely directed*, not *random*? But what about in non-directed Nature? "If we count the number of chromosomes in the members of the different species of a genus of flowering plants, we often find that the numbers are simple multiples...of a basic number...of chromosomes. The obvious interpretation is that the different species originated by hybridization followed by doubling of the chromosome numbers."[57]

Obvious interpretation? Is there not the no-small-matter of *mechanics* to consider? How, in Nature, would the (precise) doubling of chromosomes have happened in countless species following a process as typically non-productive as hybrids?

In *The Origin*, Darwin realized he had much to explain about the almost universal *sterility* of *species* when crossed over, compared with the almost universal *fertility* of *varieties* when crossed. Darwin's attempt at an answer only serves to underscore the fixity of species despite the mutability of varieties within those "fixed" species. The difference between the two is seen readily enough in the differences between their reproductive organs and functions—differences which in *varieties* can be explained by gradualism, but in "*higher species*" cannot. No amount of gradualism can provide the kind of differences in reproductive organs and functions required for upwardly-evolving, sexually-unique speciation.

In the *Descent of Man*, Darwin comes full circle back to the obvious fact of distinctive species:

> Whenever it can be shown, or rendered probable, that the forms in question have remained distinct for a long period, this becomes an argument of much weight in favor of treating them as species. Even a slight degree of sterility between any two forms when first crossed, or in their offspring, is generally considered as a decisive test of their specific distinctiveness; and their continued persistence without blending within the same area is usually accepted as sufficient evidence, either of some degree of mutual sterility, or, in the case of animals, of some mutual repugnance to pairing.[58]

Given Darwin's own test, just how convincing is the supposed evolution from one sexually-distinct species to another? At least horses and donkeys can mate and actually produce offspring.[h] The fact that such exceptions are extremely rare among millions of sexually-reproducing life forms merely underscores the obvious—that the species-specific barriers which prevent interbreeding (and thus evolutionary progression) are themselves stubborn as a mule!

Does "Evolution 2.0" Help?

Among the shelves of books I've read on the subject at hand, one of my favorites is Perry Marshall's *Evolution 2.0—Breaking the Deadlock Between Darwin and Design*. An electrical engineer with vast experience in the digital world, Marshall draws from principles of information systems and computer programming to explain the utterly fascinating inner workings of cells in biological development and evolution. Focusing particularly on *codes* as the language of information—be it computer code or DNA code—Marshall takes "five little stones" to slay the Goliath of classic evolution's two most central features: *randomness*

---

[h] Note that the eggs and sperm are already in place. However, the production of gametes in the offspring is another matter. Sterility is inevitable unless there is some form of *mosaicism* that allows the gametes to be either horse or donkey. They cannot be a mixture.

and *gradualism*. If any argument ought to put those two sacrosanct assumptions to rout, it's Marshall's.

Suddenly, though, it's not quite the rout you think it is. Just when you're ready to award him the Nobel Prize, Marshall makes the same mistake as Charles Darwin and classic evolutionists: extrapolating wrongly from the observable to the unobservable. On the level of the observable, Marshall brilliantly describes incredible design at the cell level, and even what looks to be something very much like intelligent decision-making on the part of the lowly cell, whether individually or in cohorts. Yet, when Marshall moves beyond that insightful analysis to affirm the common descent so integral to the wider Evolution Story (without attempting any in-depth argument), he fails to take into account the crucial mechanisms—sexual and otherwise—that would have been required of even the most intelligent of cells. Common descent from primitive life forms would have necessitated a gargantuan leap from their lower level of *cell activity* to the far more complex level of *entire organisms*—the only level (as Marshall notes) upon which natural selection can possibly operate.[59] Intelligent cells notwithstanding, you simply can't get there from here.

Playing on Marshall's "Russian dolls" analogy,[60] the smallest doll in the stack of dolls may be utterly amazing, but, even if it was feverishly editing DNA codes all day long, it couldn't possibly design and produce the largest doll in the set. Nor, certainly, in millions of sets (species), each with their own unique codes for mating and reproduction. As Marshall says, "Natural selection is powerless to craft complex subtle changes, because the unit of selection is not one base pair or gene, but an entire organism."[61] Which is to say, in the context of this book, that it's not just a matter of having clever cells, but coming up with radical reproductive transitions from one entire organism to another, "higher" organism.

What Marshall's "Russian dolls" *can* tell us is that their similar design and shared features point, not to common descent, but to a common designer. A common designer who, acting intelligently and purposefully, would have had the freedom to do precisely what cells are able to do: arrange and rearrange DNA patterns as deemed beneficial. (As Marshall himself suggests, it's not unlike an author swapping whole sections between two chapters, or perhaps repeating the same paragraph in two separate books. I've done that in this book!) So, if you observe similarity in DNA code, or perhaps anatomical features between two separate species, there's no compelling reason to assume it didn't happen by design, and actually far more reason not to believe it happened randomly.

Rapidly evolving cells can do many amazing things (both constructive and destructive), but they can't do everything. Among the things that intelligent cells cannot do is simultaneously produce separate genders capable of mating and reproducing in a sexually-unique way in the first generation of a new species. Or keep a developing novel species ticking over long enough for the first-ever uterous or first-ever placenta to develop rapidly enough to be of any use. Or to radically transform a card-carrying amphibian into the first-ever *sexually-distinct-by-a mile* reptile. Of all people, Marshall should appreciate that there's no code (DNA) to enable any of those things to occur. As Marshall puts it: "It's code first, evolution second."[62] No code, no evolution. If only Marshall had taken his winsome argument to the next level: No code for gender, no gender. No code for sex, no sex. No sex, no microbe-to-man evolution.[i]

Marshall is looking square in the face of Darwin's secret sex problem when he says, "A tree is built according to the instructions in its DNA, which existed before the tree."[63] Before the tree, before the bird, before the first fish, before the first-ever blade of grass, and—most importantly—before the first-ever sexually-reproducing organism of

[i] Marshall had it momentarily in his grasp, saying, "Once genes and traits exist, the laws of genetics and sexual reproduction go to work" (p. 34). The implication? For sexual reproduction to work, there must first be sexual genes. Where would those genes have come from?

any kind, there had to be pre-existing DNA for both the male and female of each unique species. Sex requires CODE! And not just any code, but unique code, distinct code, different code. Spoiler alert: The crucial lack of DNA code from start to finish means that the popular, best-selling Evolution Story turns out at the end to be a conspiracy thriller, for sure, but as fabricated and fictional as Dan Brown's *Da Vinci Code.*

Here's an exercise for you. Ask a trillion, trillion, trillion cells working intelligently and creatively together in Evolution's Grand Think Tank to come up with DNA code for the first-ever organisms that mate sexually. Tell these scary-bright, incredibly inventive cells to use every means available to them, including transposition, horizontal gene transfer, epigenetics, symbiogenesis, and genome duplication (Marshall's "five blades of a Swiss Army Knife").[j] Give them as much or as little time as you wish.

What do you think? Is there a chance in the world that any of Marshall's five "Swiss army knife blades" (or all of them together) could produce the first-ever male/female meiotic sex from non-sex? Or the unique sexual features of each next-higher species indispensable to the romanticized Evolution Story?[k]

---

[j] As defined by Marshall...*Transposition*: Cells rearrange DNA according to precise rules. *Horizontal gene transfer*: Cells exchange DNA with other cells. *Epigenetics*: Cells switch code on and off for themselves and their progeny. *Symbiogenesis*: Cells merge and cooperate. *Genome duplication*: Sudden, radical transformations of body plans.

[k] Working as they do at very elementary levels, processes like symbiogenesis have absolutely nothing to say on the level of complex organisms. What might be true on the micro level has zero effect on the macro level where complex organisms would have to make radical leaps from one sexually-unique plant or animal to another, "higher," sexually-unique plant or animal. The concept of symbiogenesis (first articulated in *Symbiogenesis: A New Principle of Evolution*, written in 1924 by Boris Mikhaylovich Kozo-Polyansky), has now gained wide acceptance in the scientific community, yet the origin of sexual reproduction remains as much a mystery as ever. And, not surprisingly, no one is even remotely pointing to symbiogenesis when it

The closest Marshall comes to answering those questions is in his discussion of genome duplication and hybridization. Marshall acknowledges that sterile offspring is the usual outcome when two different species mate. Nevertheless, he is charmed by the thought that, since a doubling of chromosomes may have happened in rare instances (as when emmer wheats and goat grass produce bread wheat[64]), that process of hybridization could have led to speciation of all sorts. Sorry, no cigar, and not even close.

First, no one, but no one, believes that the multiple millions of unique species on earth evolved via a process of hybridization. Second, what appears to be hybridization might be nothing more than mere conjecture from DNA similarity rather than any actual link between, say, sea squirts and hagfish, which Marshall admits would have required "utterly remarkable cellular engineering, including construction of several new body parts" (especially the distinct sexual features of each species, in both male and female versions simultaneously).[65] Third, observable hybridization is mostly a lab thing. There's little or no empirical evidence that it has actually ever happened on any significant scale in Nature. Fourth, when hybrids do occur, we're only talking about spin-off varieties, not stepped-up, radically-different species.

Having powerfully exploded the myth of evolutionary progress being dependent on random mutations and eons of time (Evolution 1.0), Marshall fails to fully appreciate that, even if there is a sense in which cells act "purposely" in redesigning their codes to meet specified needs, those "intelligent decisions" don't come anywhere close to providing an evolutionary process that could even remotely be termed "teleological" with an ultimate end-goal in mind. Not even Marshall's compelling "Evolution 2.0" framework becomes truly purposeful (teleological) simply because some miniscule "Russian doll" seems at times to volitionally choose her dance partners. Just as with the Master Code required to program cells with their remarkable ability to edit code,

---

comes to supplying simultaneous on-time delivery of the first male/female pair of each novel species.

teleology comes from the top down, not the bottom up. As does design. (The importance of getting that order right is reflective of an ancient story-line...[66])

Considering that randomness is the sworn enemy of research ("scientific investigation stops as soon as you invoke it"[67]), Marshall's innovative "Evolution 2.0" model is an exhilarating shot in the arm for healthcare and technology. Yet, one could wish Marshall had recognized that, while his model is marvelously suited to explaining how "bounded evolution" works (not randomly, but "intelligently"), it couldn't possibly be a model for "unbounded evolution" from one sexually-unique species to higher and higher sexually-unique species. For that, you'd need some kind of "Evolution 3.0" which could span all the millions of otherwise-unbridgeable gaps between sexually-distinct species. On that one, put up all the prize money you want. Guaranteed, you'll never have to pay out.

Adaptation, Speciation, and Sex

In his best-selling book, *Why Evolution is True*, Jerry Coyne keenly appreciates where the crux of the evolution debate lies. "A better title for *The Origin of Species*," says Coyne, "would have been *The Origin of Adaptations*: while Darwin did figure out how and why a *single* species changes over time (largely by natural selection), he never explained how one species splits in two...We must also explain how new *species* arise. For if speciation didn't occur, there would be no biodiversity at all—only a single, long-evolved descendant of that very first species."[68]

Immediately pertinent to the thesis of this book, Coyne points out that speciation has much to do with sex. "Speciation simply means the evolution of different groups that can't interbreed—that is, groups that can't exchange genes".[69] That would be tens of millions of *individual species*! How do we know we have a different species? Sex, baby, it's all about sex! Exclusive, pin-number-protected sex by organisms that can only compatibly mate and reproduce with members of their own distinct species. As Coyne puts it, "species are distinct not merely because they look different, but because there are barriers between them that prevent interbreeding."[70] So, "if you can explain how reproductive barriers evolve, you've explained the origin of species."[71]

Coyne's own attempt to explain breeding-restrictive speciation centers on the formation of "sister species" caused by geographic isolation when, for example, a mountain or body of water comes between two groups of the same species. "Suppose these conditions were to happen...," Coyne begins. "Now imagine...," he continues his speculative hypothesis. And yet again, "Now imagine..." After all that *hard evidence*, Coyne concludes that species "are simply the inevitable result of genetic barriers that arise when spatially isolate populations evolve in different directions."[72] Interesting hypothesis. But even if that hypothetical scenario remotely explained the odd "sister species," there simply aren't enough isolating mountains, rivers, or lakes on the planet to explain the origin of tens of millions of different species.

And how, possibly, does population divergence begin to explain the supposed evolution of oaks and acorns? With animals, it's at least plausible that changing migration patterns might set up population isolation and eventual divergence. But with *trees*?

What inevitably gets lost in all the discussion of population divergence is that populations don't have sex. *Pairs have sex, not populations.* In order for there to be common descent from lower to higher species, at least one sexually-unique pair of a higher species must evolve from some population of a lower species that mates and reproduces in a radically different way. Why, you ask, must it be radically different sex and reproduction? Because it's not just a matter of incrementally evolving a single innovative feature that conceivably might happen in population divergence. Each novel *higher* species must necessarily emerge as a package deal.

So, it's not just a fully-functional penis needed for the male of the "next-higher" species which no male in the precursor population ever had, but novel sperm, a different number of chromosomes, unique mating instincts, and on and on. Not just *penis*, but *package*. Not simply some new-and-improved sperm delivery system, but an entire orchestra of microbiotic players joined together in playing the right tune for that particular species. Dare one still insist that incremental changes throughout an entire population can produce such an illogical result?

And think about the major evolutionary intersections that would have to be crossed, such as from amphibians to reptiles. What would that crucial transition obviously demand? Not just one, never-before-seen feature of a reptile, but a complete, fully-functioning male reptile and a fully-functioning female reptile, compatible for mating and reproducing as reptiles, not as amphibians. (For starters, think about that first, never-before-seen amniotic egg. What evolutionary advantage would there be for only a partial amniotic egg? And how could a partially-evolved amniotic egg possibly produce the next reptile generation that can't come into existence without everything being in place in the first generation, including most importantly an amniotic egg!)

In order for microbe-to-man evolution to work, there is no escaping the problem of reproductive continuity. As Paul Nelson puts it, "Continuity demands reproductive capability, because being able to reproduce is an absolutely necessary (i.e., essential) condition for the organism in any evolutionary lineage….[T]here is—quite literally—no failing to leave offspring in common descent. If a hypothesized transformation pathway is biologically impossible…it could not and therefore did not happen."[73]

Whether between one species and some "higher" species, or from amphibians to reptiles, no transformation pathway is more biologically impossible than the inability of natural selection to supply simultaneous delivery of the first functionally-complete, sexually-unique male/female prototypes of each novel species or phyla.

No transformational pathways, no microbe-to-man common descent.

Model Versus Mechanism

Anyone insisting that a major transition might have happened incrementally from a *population* of amphibians to the first-ever reptile pair is allowing the Evolution model—simply as a sacrosanct model—to trump the lack of any possible evolutionary mechanism. Trouble is, *viable models require viable mechanisms.*

Say you trust the Evolution model no matter what because of its vaunted predictability? Even the slightest critical thinking turns that predictability on its head. When evolution's mechanism of gradual incrementalism cannot possibly advance the ball across something as important as the amphibian-reptile boundary, what can we safely predict about the viability of the Evolution model itself? That the Evolution model is seriously flawed. *Flawed mechanism, flawed model.* Well, at least the flawed model of "unbounded" evolution, not the model of "bounded" evolution so beneficial to scientific, technological, and healthcare research.

Jettisoning a super-model that can't possibly explain *microbe-to-man evolution* as it boldly claims doesn't in the least threaten a less pretentious model of evolution that helps us to understand *the microbiotics that affect man* (such as how bacteria cells evolve at breakneck speed). While the latter use of evolution is true science, the Grand Theory of Evolution is little more than sophisticated speculation masquerading as science.

"Not true!" insists a top-class medical researcher who was gracious enough to critique my manuscript. "The relatedness of organisms, based on shared descent, is the basis for all comparative biochemistry, which, for one thing, gives us the ability to choose appropriate non-human model systems so that we can successfully examine human diseases without experimenting directly on humans." Here, in a nutshell, is a good example of how deeply-entrenched evolutionary assumptions manage to crowd their way into an otherwise unremarkable discussion. Is the relatedness of organisms dependent on shared *evolutionary descent,* or simply on shared *characteristics*? Do we really need to know the supposed (mostly speculative) evolutionary history of two species, or simply that they have sufficiently similar systems so as to permit effective and beneficial experimentation?

For anyone concerned that giving up the Grand Theory means having to give up important strides in medicine, here's the question that needs asking: Is any current research or medical breakthrough dependent on whether evolution ever had the capacity to bridge between amphibians and reptiles, or to explain the origin of sex, or to provide on-time delivery of the first pair of millions of sexually-unique species? As alluded to in the Introduction, evolution-informed scientific research, technological advances, and important healthcare discoveries are not in the least dependent on the romanticized Evolution Story.

What Extinction Tells Us About Origins

The problem is not genuine evolution science, but the popularized story. Yet there is one aspect of the Evolution Story that is not just true, but—by some irony—is helpful in debunking that very story! Strangely enough, Evolution's much-talked-about extinction helps us to understand that evolution could not possibly have produced disparate, yet compatible, genders for each and every one of tens of millions of sexually-distinct species. Consider most especially black rhinos, blue-throated macaws, the brown spider monkey, the California condor, and the Sumatran tiger—all of which are now critically endangered, but none of which even would have existed if left to natural selection.

And herein lies the tale. When will these species become extinct? When it gets down to the last surviving male or female of the species. Right? One without the other will not alone be able to produce another generation. And at that point, the DNA of the species will be forever lost.

Does this not tell us something important about the *first appearance* of that species—that a male brown spider monkey without a fully-compatible female brown spider monkey would never alone have produced offspring that eventually could become extinct? Nor could a male Sumatran tiger with only a partially-evolved penis ever have produced the next generation of Sumatran tigers, which after thousands of generations might soon become extinct.

Interesting, isn't it? The lesson of extinction is the lesson of existence: *Survival to the next generation is always dependent on the existence of a perfectly matched pair of progenitors, whether at the dying end of a species or—even more important—at its beginning.*

Recap

- Speciation in Nature along a horizontal plane clearly happens, but the evolution of mere spin-off varieties doesn't move us along the supposed vertical chain of evolution from a "lower" species to a "higher" one.

- Wherever you look in Nature, there are constraints at the level of mechanism which, like a rubber band stretched by degrees, can only stretch so far without finally breaking.

- All sexual processes require the prior existence of unique code (DNA information), without which there can be no transitional pathway from one unique form of sex to another.

- Hybridization could not have provided the mechanism for microbe-to-man evolution, since it's mostly a lab thing (only rarely and ineffectively happening in Nature), and could not possibly explain the origin of more than a relative few of the millions of species on the planet.

- Just as a species becomes extinct when the last remaining form of that species lacks an available sexual partner, so, too, no

> sexually-unique "higher species" ever could have evolved in the first place, since natural selection could not possibly have supplied both the first male and first female simultaneously.

Question: Does it make sense to you that, for all its fascinating impact on Nature, the process of natural selection has limitations beyond which it cannot possibly operate?

CHAPTER 3

Not Simply Complex, But Unique

All cases are unique and very similar to others.
—T. S. Eliot

Looking especially at bizarre mating and reproductive habits which demand, not simply the "irreducible complexity" championed by proponents of Intelligent Design, but unique, exclusive differences in the sexual processes of distinct species, precluding any notion of common descent.

One of the more vigorous arguments offered to refute evolution (primarily by advocates of Intelligent Design) is so-called "irreducible complexity." The idea was most notably put forward by Michael Behe in his widely-discussed book, *Darwin's Black Box*, wherein Behe explains:

> By irreducibly complex I mean a single system composed of several well-matched, interacting parts that contribute to the basic function, wherein the removal of any one of the parts causes the system to effectively cease functioning. An irreducibly complex system cannot be produced directly (that is, by continuously improving the initial function, which continues to work by the same mechanism) by slight, successive modifications of a precursor system, because any precursor to an irreducibly complex system that is missing a part is by definition nonfunctional. An irreducibly complex

> biological system, if there is such a thing, would be a powerful challenge to Darwinian evolution.[74]

I suppose one could say that I am making what amounts to an "irreducible complexity" argument, not only when it comes to the intricate sophistication of the organs for sexual mating and reproduction, but also in my "half a penis, half a vagina" argument that sex could not have come about by slight, successive modifications of a precursor system. Sex truly *is* complex, and far too complex to have happened incrementally. It's *irreducibly complex*, if you wish to put it that way. But those books have already been written, and evolutionists have responded with counter-arguments of their own.

Hopefully by now you have sensed that what I'm attempting to show in this book is not simply that sexual organs and reproductive processes are complex, but—far more important—that they are *unique.* It virtually goes without saying that (particularly at the microscopic level) *male* forms and *female* forms are breathtakingly complex. With sex, however, complexity is not the central issue—whether irreducible or reducible—but the very existence of two unique units capable of contributing their mutual genes to forming a completely new species.

What neither evolution nor irreducible complexity alone can provide is the volition necessarily required on the part of two separate entities to join together in sexual union. Whether mating is driven by sexual passion (as in humans) or mere rudimentary reproductive instinct, the complexity of any single entity is not the whole story.

Sex—A Dickens of a Curiosity Shop

In the previous two chapters, we've pointed out that, among the millions of sexually-reproducing life forms, natural constraints prohibit variation beyond a breaking point of interbreeding species. We noted that, despite all the variations which might be induced by either Nature or man, any truly new species would have to include radically different sets of progenitors—namely, a fully-functioning male and a fully-functioning female—both obviously compatible with each other, no matter how weird they might be. Which takes us to countless life forms that reproduce in unique, often bizarre ways. *Complexity alone is not the issue, but uniqueness.*

Endless volumes could be written on the utterly fascinating idiosyncrasies of mating and sexual reproduction among the millions of species in Nature. You may already know, for example, that when dogs copulate, the distal portion of the penis swells markedly so that it becomes firmly lodged within the vagina. Such "genital locks" are rare among the primates. Certainly humans (along with monkeys and apes) have no lock on sex![75] The questions for us are two. First, how did the prototype "doggie couple" originate from some species clearly *not* dog, and not mating like dogs mate? Second, if genital locks are a necessary part of the dog-mating repertoire, how did natural selection develop the "genital lock" gradually over *successive generations* when *no generation* could have come into being without that necessary feature?

There are also the strange solid, rubbery "copulatory plugs" found in the chimpanzee, where coagulation produces a soft, whitish gelatinous material that hardens.[76] If "copulatory plugs" are integral to chimp sex, how did the very first chimp couple copulate (and reproduce) with only partially-developed plugs?

Surely by now you're picking up on the pattern. Just how many separate hypotheses does it take to explain what formidable odds evolutionary gradualism faces among the millions of different species, *each with its own unique sexual oddities*? Does natural selection explain *every* distinctive phenomenon, or is it simply assumed that—since the Grand Theory is beyond challenge—every distinctive phenomenon *must* have an evolutionary explanation?

The farther down you drill from macro to micro, the more complicated sex gets. As Alan Dixson describes it for us, "Some mammals have relatively long and convoluted oviducts, whereas in others the ducts are quite short. There are differences also in the morphology of the ciliated fimbria (which guide the ova into the entrance of the oviduct) as well as in internal features such as the structure of the uterotubal junction."[77]

Here again is a reminder of the behind-the-scenes "supporting cast" required by the more recognizable bits of sexual anatomy. Before a particular species has either long or short oviducts, the first intact prototype of that species must have possessed all of the supporting equipment, fully formed and functioning. Evolutionary gradualism is simply hopeless to provide on-time delivery for each and every interdependent part required for the job—*especially for brand-new customers who insist on doing things differently from everybody else.*

Although wings are not part of a bird's sexual anatomy, the supposed evolutionary development of wings clearly demonstrates the problems of gradualism when it comes to complex structures such as a bird's mating and reproductive systems. As Denton observes:

> Take away the exquisite coadaption of the components, take away the coadaption of the hooks and barbules, take away the precisely parallel arrangement of the barbs on the shaft and all that is left is a soft pliable structure utterly unsuitable to form the basis of a stiff impervious aerofoil. The stiff impervious property of the feather...depends basically on such a highly involved and unique system of coadapted components that it seems impossible that any transitional feather-like structure could possess even to a slight degree the crucial properties.[78]

Evolutionist though he was, Stephen Jay Gould poked fun at the notion of evolutionary gradualism, using rather more colorful language than some:

> But how can a series of reasonable intermediate forms be constructed? Of what value could the first tiny step toward an eye be to its possessor? The dung-mimicking insect is well protected, but can there be any edge in looking only 5 percent like a turd?[79]

Having half the required sexual and reproductive organs necessary for a species is far more problematic than some creature with only 5 percent of its protective camouflage or the ever-troublesome half-a-wing. If evolution has provided only half the required sexual and reproductive organs necessary for a given species, you never get to Generation One or Generation Two in order for evolution to develop those marvelous flying wings, much less some "next higher" species reproducing in a radically different way.

Again and again, it's the same problem, whether it's half a bird's wing or half a penis or vagina. When it comes to sexual reproduction, you just can't do it half way! *Nor (for any new, sexually-unique species) the same old way!*

An Acorn of Truth

If we focus too much on the animal kingdom, we might miss one of the best illustrations of how sexual development could never have happened by gradual evolution. Take, for example, the mighty oak (*Quercus spp.*), of which there are some 500-600 species (varieties). Oak trees are monoecious, meaning that each tree produces male and female flowers. The male's pollen is carried by insects or wind to the female flowers on neighboring oak trees. After fertilization, the oak produces its trademark acorns *which no other tree produces.*[1]

---

[1] If you say that acorns might have evolved from similar-appearing beechnuts or chestnuts, then go back to whichever one you think evolved first and explain how the first *pair* of non-oak trees evolved to propagate its own unique nut, much less how it then morphed into the acorn.

So, let's talk about the very first acorn ever to appear on the face of the earth. Which came first, the acorn or the oak? If you say the acorn, what possibly would be its evolutionary precursor, since the acorn is unique to the oak? And if you say the oak tree itself, what possibly would be its evolutionary precursor, since oak trees are the unique product of acorns?[m]

And don't overlook this crucial distinction: When we're talking about the 500-plus species of oak trees throughout the world, no dramatic sex change is necessary. All 500-plus varieties have essentially the same mating features, and all produce those trademark oak acorns. So, drawing from the old line about ducks, if it has leaves like an oak, mates like an oak, and produces acorns like an oak, it *must be* an oak! The particular species of oak hardly matters.

By contrast, there would be nothing simple about evolving the very first oak from any of it's closest non-acorn-producing neighbors in the forest. Even what might seem only a slight modification would require a *package of changes*. As Michael Denton explains: "Any change which on the surface may at first appear quite trivial, on closer examination would inevitably necessitate extensive reorganization of the entire anatomy and physiology of the organism."[80]

To get oak trees from non-oak trees, it's not simply a matter of some aberrant mutational cross-dressing, but a shocking genetic sex-change. Beginning with never-before-seen "oak DNA," the problem is coming up with the first *two* oak trees, and then—wonder of wonders—the first-ever acorn. Achieving that result would require a *whole package of changes*—morphological, physiological, even behavioral.

[m] In the crystal-clear case of oak and acorns, would anyone have the chutzpah to offer the usual lame explanation: divergence of some precursor population as an explanation?

At Major Intersections, It's All, or Nothing at All

And to think that the problem of producing that single, sexually-unique tree is miniscule compared with the major transitions required for climbing the so-called "evolutionary tree." Merely consider the supposed evolution of reptiles from amphibians, described by Mark Jerome Walters from an evolutionary perspective:

> When animals left the sea...they were deprived of the wet environment required by egg and sperm. So instead of depositing sperm and egg externally in the water, many land animals adopted a means of internal fertilization, depositing gametes in a sequestered internal environment. Reptiles encase the fertilized egg in the watery interior of a shell.
>
> The move from sea to land had great implications for courtship. Whereas in the sea, partners merely had to be brought into proximity with each other, on land actual mating was usually required. This meant major changes not just in courtship patterns, but also in the physical design of the animals themselves. A simple vent for expulsion of egg or sperm would no longer do.[81]

Adopted a means of internal fertilization, you say? Easy for some! Could we please roll the tape of evolutionary history back to the point where there is not a single reptile on the planet, only amphibians? In order to have Generation One of the very first reptile, what all do we need? For starters, Walters says we need more than a simple vent for expulsion of egg or sperm. By what evolutionary process is that going to happen? Just how many permutations would it take to develop from scratch something like the reptile penis; and how many permutations would it take to simultaneously develop the very first (entirely unique) amniotic egg? Most important of all, what evolutionary progress can possibly be happening in the meantime with only a partial penis and a half-developed egg (not to mention all the other organs, structure, and microbiotics necessarily required for a fully-functioning reptile)?

Remember, too, that all this first-time-ever transitional evolution would have to take place in the rollicking sea, where amphibian eggs and sperm are floating about randomly, with the probabilities of an actual fertilization already incredibly slim. And then there must be some kind of unexpected mutation leading to the first tiny change that might start the ball rolling; but the ball can only keep rolling if both kinds of reptile-like sexual apparatus keep occurring simultaneously...in that random, rollicking sea! And finally, we need to have both reptile sperm and reptile eggs fully developed before there is any chance of moving on to Generation Two. Richard Dawkins' "selfish genes" can't pull off this logic-defying feat, nor massive DNA mutations, nor evolutionary gradualism, nor "coevolution," nor survival of the fittest, nor sexual selection, nor sheer fanciful wizardry!

So, which came first: the fully-formed reptile or the amniotic egg? Surely, it must be the reptile. There's not a chance in the world of random chance that an amphibian ever could have produced a reptile egg! As Michael Denton reminds us, "The amniotic egg of the reptile is vastly more complex and utterly different to that of an amphibian. There are hardly two eggs in the whole animal kingdom which differ more fundamentally."[82] Considering that it would take at least eight quite-different innovations, how possibly could the amniotic egg of the reptile come about gradually as a result of a successive accumulation of small changes?

If the reptile came first, not the reptile's marvelous amniotic egg, the question remains: How possibly could the first reptile capable of producing the first amniotic egg ever have come into existence by a process of gradual change from the radically-different reproductive system of amphibians? Until a land animal itself *comes into being* (in Generation One), it can't possibly adopt a means of internal fertilization (to produce Generation Two). Darwin's Grand Theory of evolutionary progression from the sea to the land, from amphibian to reptiles, is romantic at best and, at worst, sheer scientific fraud. When it comes to major intersections from one archetypical form of sexual reproduction to another radically-distinct archetypical form of sexual reproduction, without everything that is *sexually unique*, you have nothing.

The same goes for Carl Zimmer's nonchalant assumption that "About 140 million years ago, mammal evolution produced two branches that would turn out to be the most successful of all. One was the marsupials, which include living animals such as the kangaroo, the opossum, and the koala.... A fertilized marsupial egg does not develop a shell; instead, the embryo develops for a few weeks until it is the size of a rice grain, and then it crawls out of the uterus" making its way into a pouch on its mother's belly.[83] The other branch, supposedly leading to us humans, is known as the placentals, whose babies stay in the uterus until they are much larger—made possible by having a placenta surrounding the embryo.

The obvious question is: How did evolution ensure that in the very first generation of the marsupial from whatever might have been its non-marsupial progenitor it would have been possible to have new sexual organs for marsupial mating and reproduction, including the unique marsupial egg having no shell? Compounding the problem, how, from the same common ancestor, could placentals possibly have developed that all-important placenta to nourish the developing fetus? Could either of these have happened *gradually*? Half a penis; half an egg; half a placenta?

In his introduction to subsequent editions of *The Origin*, Darwin observed that "even slight advantages, of one-half of one per cent or less could have important evolutionary effects."[84] While that might be true in terms of wing variations or color changes, it can't possibly be true in terms of novel changes in sexual reproduction. If half a penis and half a vagina are problematic, how much more so one-half of one percent of a penis or a vagina!

Nor should we overlook the totally-ignored "switching problem." Even supposing some evolving amphibian slowly acquired the characteristics of a reptile, at some snapshot moment it would have to *turn off* all its amphibious sex routines. How could any amphibian produce the first generation of reptiles while it remains essentially amphibious? When it comes to some supposed transition from one major phyla to another (or, for that matter, even from one species to another), distinct methods of reproduction can't simultaneously co-exist!

To have anything, you've got to have everything. The right everything, not just any thing.

Other Fascinating Case Studies

Try to imagine the "right kind of everything" you'd have to come up with simultaneously (starting, yet again, with both *male* and *female*) for the incredible mating flight of the dragonfly:

> The male flies ahead of the female and grips her head with terminal claspers. The female then bends her abdomen forward and receives the sperm from a special copulatory organ which is situated toward the front on the under-surface of the abdomen of the male dragonfly and which he fills with semen from the true reproductive aperture before the start of the mating flight. This strange manoeuvre, which seems a curiously roundabout way to bring sperm to egg, depends on the unique and complex machinery which forms the male copulatory organ.[85]

The mating flight of the dragonfly is not just fascinating, it's telling. As R. J. Tillyard pointed out in his book *The Biology of the Dragonfly*, "... the apparatus of the male Dragonfly is not homologous with any known organ in the Animal Kingdom; it is not derived from any pre-existing organ; and its origin, therefore, is as complete a mystery as it well could be."[86] Translation? In its complexity, oddity, and uniqueness, the simple dragonfly defies all evolutionary logic.

Same with common, ordinary fleas, which, when you scratch beneath the surface, you discover a reproductive system even more complex than that of the dragonfly, and without parallel in any other order of insect.[87] Fleas, *sexual wonders*? Go figure!

If you're looking for sexual wonders, you can hardly beat the Brazilian cave creature known as *Neotrogla*, a flea-sized winged insect which combines both sexual role reversal (where the female is the promiscuous aggressor, as with the scorpion fly) and an anatomical reversal (with the male being penetrated by the female "penis," not unlike seahorses). Giving new meaning to "hooking up," there's even a genital lock like dogs, but with spiky-spined grappling hooks for marathon sex sessions lasting over a period of 40-70 hours![88]

With *Neotrogla*, the mechanics of sex may be reversed (as if a fighter jet with a fixed boom were refueling a flying tanker instead of the usual way around), but it's still the female receiving sperm from the male. Attempts to explain *why* there might have been an anatomical reversal are interesting, but don't begin to explain *how* such mutually-compatible anatomy might possibly have evolved "in tandem" for the first generation of *Neotrogla*.

Sex doesn't get more bizarre than with the *praying mantis*. (Guys, you'll never have sex again with both eyes closed!) After the male mounts the female, clasps her sides with his forelegs, and rocks his abdomen up and down, it takes about half an hour before his sperm is transferred to the lovely lady...who (perhaps spurred on by his antennae violently beating her) suddenly twists around and bites off her partner's head! Relieved of his brain, the male loses all his inhibitions and begins rocking his decapitated body like there's no tomorrow (which, for him, there isn't!).

Certainly solves the problem of potential unfaithfulness by one's sexual partner, but it doesn't solve the evolutionary problem of how that unique physical apparatus and those corresponding mating habits ever came to be in Generation One so as to progress to Generation Two. From what closest living being would the *praying mantis* have emerged? As always, how did we get both the (unsuspecting) male and the (cannibalistic) female developing *simultaneously*—each having complete bodily form and full sexual functioning, including the bizarre cannibalism bit.

Think that's weird? What blind forces of gradual evolution could possibly explain *the alternative sex life* of the praying mantis? In some cases, the female may actually ambush the male and eat his head *before* he ever mounts her, in which case automated, virtually robotic, movement in the male's legs hauls his body on top of the female whereupon it frenetically has sex with the femme fatale who has just bitten off his head![89] What, possibly, could pre-program the male's legs to act independently of the brain so as to mount the female with his decapitated body?

Sexual Organs and Ripley's Believe It or Not

Reporting for the BBC in an article intriguingly titled, "The Twisted World of Sexual Organs," Colin Barras provides us with an incredible description of some of the world's most bizarre sexual organs.[90] Barras begins the article with a bit of TMI, describing how a Korean woman eating squid had triggered a release of the dead male squid's spermatophores, which in the normal mating process act as tiny guided missiles to implant the payload of sperm into the soft tissue of the female squid near her genital opening, releasing sperm as the female lays her eggs. In the case of the Korean woman, the target ended up being her tongue, cheeks and gums! (Squid tonight, anyone?)

Barras then takes us on a wild ride, noting generally about insect genitalia that:

> There is a baffling level of variety in their shape and size. So much variety, in fact, that some closely related species that otherwise look identical can be distinguished purely because of differences in the shape of their sexual organs. This doesn't just apply to insects. A quick look at the sexual

> organs can help distinguish between closely related species of reptiles and mammals too. Many male mammals—although not humans—carry a bone in their penis. The variation in shape and size of this penis bone is astonishing, if a little eye watering from the female perspective.

There's more. Since flatworms are hermaphrodites carrying both male and female genitalia, a "fencing match" ensues in which "each flatworm attempts to stab the other with its two-pronged, fork-like penis. Sexual intercourse ends when one flatworm registers a successful hit on its opponent and transfers a packet of sperm." Nico Michiels of the University of Tuebingen, Germany, suggests that flatworms have a strong motive to hone their fencing skills since the winner is free to roam while the loser has to "stay home and tend the babies." Even if that explained the *why* of flatworm sex, it doesn't begin to explain the *how.*

And how about the male damselfly, with a penis whose tip carries two horn-like structures, each coated in tiny spines for brushing away any rival sperm from the female's reproductive tract. Or the male weevil's penis that has spines which sink into the lining of the female's reproductive tract, causing her injury. Might that have the benefit of discouraging the female from mating with a rival in the near future, Barras asks?

"You dirty rat, you!" might well be applicable to the Norway rat that hooks together hundreds of little fellas with tails to form a mega-sperm that can power its way towards the eggs faster than a single sperm. Or perhaps even more so those other rats that have "kamikaze" sperm which are thought to form a tangled mess to ensnare the sperm of rival males. Rats!

And then there's the Argentine lake drake that, in the blink of an eye "explosively releases his 40-centimetre-long [16-inch] penis, using it to lasso the female and force himself on her." One surely has to ask just how beneficial an adaptation (to be passed on to the next generation) would, say, a ½-inch lasso be? And from whence would come the biological instinct to use even a 16-incher? What possible evolutionary process would have kicked in at just the right time to provide the enzymes to enable the movement of the lasso, or to supply the complete set of biological mechanics necessary to simultaneously bring together all those complementarian functions?

Barras' bottom-line conclusion is telling: "Perhaps the ultimate truth is that the world of sexual organs is such a strange place because... animal genitalia evolved to make both love and war." That clever but meaningless closure comes nowhere near fulfilling Barras' promising sub-title: "How evolution solved the problem of conception." Nothing more is being said than that, since evolution explains everything (the given operating premise), then (conclusion) evolution explains everything! By that reasoning, even tautology is the end product of evolution!

The true "problem of conception" is that, in millions of weird and wonderful different species, there must have been—from the very first time onwards—a fully-developed male and a fully-developed female, with all the bells, whistles, compatible sexual organs, microbiotic reproductive cells, precise DNA, and morphological capacity for the conception and development of the weird and wonderful offspring unique to that species. That's a tall order indeed, and a *package deal* that evolution's gradualism could not possibly have supplied.

Whether for making love or making war, the burning question remains: HOW? How possibly could evolution have come up with all these utterly fascinating sexual organs—fully-formed, and in tandem per species; and with all the required support systems necessary for the sexual organs to work as they do? How did they appear suddenly from some progenitor *not* having those unique, highly-complicated sexual organs?

If you insist on the felicity of limitless time, how could these strange organs have been sufficiently functional if only partially evolved at each minor stage of progress? (Don't forget the uselessness of a ½ -inch lasso.) The fact is that the problem of conception is not in the least explained by evolution. To the contrary, evolution's requirement of minor incremental changes over vast periods of time is the strongest proof yet of a fundamentally flawed theory.

As these (and countless other) bizarre forms of sex confirm, when it comes to the diversity of sex throughout the species, evolution's prized gradualism is a myth of mythic proportions.

Indeed, the more radical, specialized, and mind-boggling the mating couple, the less likely gradualism could have produced such creatures with their bizarre sexual oddities.

Estrus—Yet Another Unexplained Sexual Disparity

There is no end to all the sexual curiosities which are assumed, but left wholly unexplained, by Darwin's Grand Theory. Consider, for instance, that—unlike most animals—the human species is sexually active all year round and throughout the menstrual cycle.[91] That is not generally the case with the animal kingdom, in which estrus is common. The term *estrus* derives from the Greek word for gadfly, whose bite drives animals into a frenzy, and thus aptly describes what commonly is known as being "in heat." In most animals, it is only during the ovulatory phase of the sexual cycle that the female is receptive to the male's sexual overtures.[92]

Consistent with Nature's fixed boundaries of mating and sexual reproduction, the estrous cycle varies from species to species. "*Mono*estrous species," (including the fox, the bear, and the wolf) breed only during a single season each year. "*Di*estrous species" (like dogs) typically have two breeding seasons a year (though it can vary from one to three). "*Poly*estrous species" (like cows, cats, and domestic pigs) can be "in heat" several times during the year; and, naturally, those randy rabbits are always ready!

Nor, of course, is it just the number of mating seasons per year that are unique to various species, but also their varying lengths of estrus. For example, the length for cats is 14-21 days; 4-13 days for dogs; 4-10 days for horses; 17 days for sheep; 21 days for cows, pigs, and goats; 23 days for donkeys; and 16 *weeks* for elephants! What do you think? Did all of this fixed variety come about by blind forces, or even a supposed "law" of natural selection?

Drawing from evolution theory, one can easily hypothesize (particularly with larger species) that mating seasons would generally correspond to availability of food and other environmental considerations to maximize the offspring's chances of survival. However, that still begs the question: How by any means would estrus have evolved in the female of *Generation One* of any particular species? How would that first prototype know which would be the best season for estrus? How would she develop whatever signals she sends by way of invitation to potential suitors, whether pheromones (scents) or otherwise? And how did those suitors appear simultaneously, complete with the ability to read her come-hither signals? At some point, we've simply got to slow down the whirring time machine of evolution to stop-action, slow-motion! With each "frame," there are more and more questions to ask.

Even Stephen Jay Gould saw the danger in all the high-speed hypothesizing among evolutionists. "Step way back, blur the details, and you may want to read this sequence as a tale of predictable progress: prokaryotes first, then eukaryotes, then multicellular life. But scrutinize the particulars and the comforting story collapses."[93] It's even more problematic when evolutionists speed blindly past one problem after another, including especially the problem of sex. Despite sounding an important warning that the devil is in the details, Gould himself speeds right by the Queen of evolutionary problems!

Speaking of details, since human females don't generally experience estrus, anthropologists have asked why there was a *loss* of estrus during human evolution. Here's an even better question: Is it reasonable to believe that estrus in the animal kingdom ever first came about by evolutionary gradualism?

And then there's the best question of all: *Is it remotely possible that evolution's Grand Theory itself is the problem, not the myriad problematic phenomena that are inconsistent with it?*

Distinct, Yet Compatible Males and Females

A somewhat different type of oddity deserves at least passing mention. How possibly does evolution explain the secondary levels of *sexual dimorphism* so ubiquitous among sexually-reproducing beings—that is, the marked differences in size, or perhaps in coloration, between the male and female of any given species? How possibly could that dimorphism have happened in each newly-developing species?

And how—despite typical dimorphism between males and females—does evolution explain the uncanny compatibility between the (often bizarre) mating and reproductive processes of the male and the female in each unique species of living beings? Despite the disparity in body sizes, quite unbelievably the penises and vaginas actually match—and (regardless of their particular kind of sex organs) *must have matched perfectly* in the first prototype pair of each and every species.

If You're Looking for a Good Book...

When it comes to cataloguing the fascinating world of sex from unique species to crazy-unique species, have I got a book for you! Run, don't walk, to get your hands on a copy of *Dr. Tatiana's Sex Advice to All Creation*, by Olivia Judson (Imperial College London), who just might know more about sex in Nature than anyone. Cleverly written as an advice to the lovelorn column for all creatures great and small, this rollicking, saucy tome has been hailed by reviewers as a scientific tour de force. By any measure, Judson's book is an encyclopedic description of the fascinating, often bizarre mating and reproductive habits of countless plant and animal species. No book could better emphasize the point of this chapter, that sex in Nature is not only amazingly complex, but (far more important) unique, idiosyncratic, and—apart from some occasional frisky (non-reproductive) hanky-panky—reserved exclusively to one's own species. As Judson reminds us, the working definition of a species is a group of organisms capable of interbreeding.[94] Others need not apply.

Judson's book is also worth a read if for no other reason than to confirm that even the best and brightest evolutionists consistently turn a deaf ear to the problem of evolutionary sex virtually screaming in our face. Judson's encyclopedia of weird and wonderful sex is a perfect illustration of the principle that description is not explanation. From cover to cover, Judson simply assumes that, no matter how bizarre or mind-boggling the sex, surely, it all came about by evolution. Given the vast array of perfectly-paired, species-distinctive sex, that's some leap of faith, reminding me of Judson's great line: "It is a fascinating idea but extraordinary claims require extraordinary evidence." Surely, the corollary is: extraordinary sex requires extraordinary explanations.

For Judson, the most fundamental conundrum—how evolution might have come up with sex in the first place—is literally an afterthought. In her two-page Postscript, Judson finally asks, "How did sex begin?" but then—along with everybody else—promptly changes the question to "*why* sex?" (which might explain the evolutionary advantages of sex, but deftly avoids having to come up with a credible process by which sex first came to be). As for the more crucial *how* question, Judson punts, saying,

> However speculative the origins of bacterial sex, the ideas look like a mighty edifice compared with how little we know

> about the origins of the sort of sex that humans, birds, bees, fleas, green algae, and other eukaryotes conduct. Remember: eukaryotic sex is a complicated process requiring that each parent donate one complete set of his or her genes. Probably it evolved only once. But exactly how or why is a deep mystery.[95]

I appreciate that it wasn't Judson's brief to explore the origin of sex. But, if not the origin of sex, one could at least hope for some passing mention of the next most obvious question raised by what she did write: How possibly could the natural selection that so marvelously produced all this incredible sexual activity in Nature even remotely have provided simultaneous on-time delivery of the first-ever male and female of each of the sexually-distinct species Judson cites in her book?

In her informative and entertaining page-turner, Judson regales us with all sorts of fascinating scenarios, from the battle of the sexes (sometimes including brute force and even rape), to multiple genders (up to 500!), to relatively rare monogamous pairings in animals, to anomalous female-dominated reproduction, to those dual-gendered hermaphrodites, to unseemly incest, rampant promiscuity, and examples of homosexual dalliance.

It's virtually everything you ever wanted to know about plant and animal sex, but were afraid to ask...except for the most important thing of all: how the uniquely-matched male and female forms of each of millions of species could have evolved in synchronized harmony—with each male/female pair (virtually always distinct from one another in size, shape and patterns of behavior) being perfectly compatible for mating and reproduction from day one. Disappointingly, there's not the slightest hint as to how blind forces of evolution could ever have pulled off that improbable feat in even a single species, much less in millions upon millions of species.

Indeed, when it comes to evolution's bedrock theory of common descent, Judson doesn't offer even a modicum of speculation as to how the uniquely-matched male and female of species X could ever have given rise to the necessarily-tandem evolution of the first uniquely-matched male and female of the next higher species, Y—even given all the mutations, genetic variables, and environmental factors you wish to assume. (That's *higher species*, not simply variations on theme where "new species" are nothing more than spin-off varieties of the same basic organism, whether fruit flies, sea urchins, or abalone.)

Nor does Judson shed any light on how only partially-evolved intermediate generations from species X (or from some spin-off variety of species X) possibly could have produced any offspring at all. Much less so the very first male and female forms of species Y (a distinctly higher species) required for reproducing the species-Y offspring necessary for the subsequent evolution of yet higher and higher species on the evolutionary ladder.

Aren't these the crucial questions you would expect to be answered when the front cover of the book promises to deliver "The Definitive Guide to the Evolutionary Biology of Sex"? Evolution literature never fails to astound. Why are these make-or-break questions virtually never asked, much less answered?

If the scientific community could ever bring itself to ask the right questions, it would soon be clear that (as Judson says so deliciously curt regarding Bateman's principle[n]) *microbe-to-man evolution has a fundamental flaw: it's wrong.*[96]

---

[n] According to A. J. Bateman (1919-1996), since females invest more in the reproductive process than males, males are fundamentally promiscuous, while females are fundamentally selective, or chaste. Not necessarily so, says Judson, citing case after case where females are notoriously promiscuous.

Recap

- Sexual organs and reproductive processes are indeed complex ("irreducibly complex," if you wish), but—far more important—they are *unique* to each species.

- When moving from one unique species to another unique species, a whole new package of mating and reproductive mechanisms must be in place simultaneously in both the male and female forms; and, obviously, they must be compatible.

- When it comes to major intersections between one archetypical form of sexual reproduction and another distinct archetypical form of sexual reproduction (say, from amphibians to reptiles), the required package of internal and external sexual changes is simply too radical to be explained plausibly by natural selection.

- The incredible array of bizarre sexual oddities found throughout Nature is irrefutable proof that no evolutionary transitions possibly could have produced them.

- The glaring problem these millions of "missing sex links" pose for the notion of common descent is not even addressed by the scientific community, much less given an attempted explanation.

Question: Considering what you've been reading about the bizarre sex life of millions of Nature's endlessly-fascinating creatures, can you see why natural selection would be hard-pressed to deliver the first-ever compatible male/female pair for all those sexual marvels?

PART TWO

The Gap Evolution Couldn't Possibly Jump

CHAPTER 4

Incredible, Mind-boggling Sex!

Most human beings have an absolute and infinite capacity for taking things for granted.
—Aldous Huxley

Laying a foundation for the book's main thesis by exploring the intricacies of human reproduction and, drawing from that, focusing on the complexity and uniqueness of sexual reproduction at even the simplest possible level.

For a generation that's all about sex, it's amazing how little we know about sex. Certainly, we know about penises and vaginas and something of what might happen over the ensuing nine months, but we're virtually clueless about what happens behind the scenes to make all of that happen. And that's just in humans. What about sex in chimps, iguanas, petunias, and peas? Or in fish, or grass, or spiders and fleas?

Sometimes we just take too much for granted. It's certainly true of sex, but more importantly it's true of theories *about* sex. In fact, it may be that our taking too much for granted about sex allows us to take far too much for granted when it comes to the origin of sex.

So it's time to have "the talk"…you know...about the birds and the bees, and, oh yes, about mommy and daddy. As you might guess, this will be a much more serious talk than usual. We'll be digging deep into the backstory of sex, without which sex isn't sex—not in bed, not in reproduction, not in theory, not in history, not in fact.

I've tried to make this chapter as newcomer-friendly as possible, so if this is all-new territory for you, hang in there as best you can and don't get put off too much by the details. Crucially, it's the details that so fully expose the Queen of evolutionary problems, and, simply for their own sake, the details that are so incredibly amazing!

Did you know, for instance, that the average number of spermatozoa (sperm) in a single human ejaculation is 236 million![97] Or that only relatively few of the male gametes (sex cells) actually come anywhere close to "home base"?

Did you know that only 5% of semen is made up of fluids contributed by the testes? Or that the sperm which leave the testicles and move faster than a speeding bullet through the *vas deferens* tube leading to the penis, are mixed with secretions from a number of different reproductive glands which provide most of the ejaculate—including 60% from the seminal vesicles and 30% from secretions in the prostate?[98]

Did you know that in the process of the journey from testicles to penis the sperm spend time "awaiting bail in a holding cell" (the *epididymis*), morphing biochemically in order to have increased mobility and enhanced capacity to fertilize?[99] Or that the sperm's breakneck ride through the *vas deferens* is made possible by wavelike contractions in its three-layered muscular walls? Or that, as they're speeding along, the sperm get "car-washed" with various beneficial substances including a special alkaline secretion that helps the sperm survive in the hostile acidic environment of the vagina? Go figure! Other secretions contain an impressive variety of substances including sugars (notably fructose) for energy; mucus to help the sperm become better swimmers; a secret ingredient (*prostaglandins*) to stimulate uterine contraction; and proteins, which enable the coagulation of the semen.[100]

And would you believe that, as the millions of spermatozoa then begin to accelerate through the urethra (the passageway normally used for urination, but temporarily commandeered by sperm on their way to ejaculation), there's even a nifty little fluid splashed into the urethra to prevent any residual urine from contaminating the sperm? No detail has gone unnoticed.

Then there's that perilous post-ejaculation journey to consider. In order to reach the coveted "lost ark" (the ovum), the spermatozoa must make like Indiana Jones and maneuver their way through dangerous anatomical and physiological minefields in the vagina, cervix, uterus, uterotubal junction, and the oviduct, without faltering at any hair-raising point.[101] Yet despite the mad race to the ovum, the sperm actually hang out momentarily in the isthmus of the oviduct (fallopian tube) before migrating to the upper part of the oviduct (the ampulla) where fertilization takes place.[102] Why the unexpected layover? To create a temporary pool of gametes and to rev up their engines in anticipation of the final lap; but most of all to wait for ovulation to occur and for the ovum to be transported to the ampulla.

If you think all of that complicated maneuvering in the male is impressive, the female's sexual reproductive structure is equally incredible. Merely consider the daunting trip the egg must take to meet the sperm for their big moment. The ovary, of course, is "the egg basket" from which the egg begins its journey. (Actually, there are two ovaries, one on each side of the body.) Next up is the oviduct, or fallopian tube, through which the egg eventually travels to the uterus. But it's in the oviduct along the way that the egg meets the winner of the sperm race. This "man of the match" male gamete curiously mimics the male penetrating the female in intercourse by itself penetrating the egg, thereby fertilizing it. (What are the odds of such a curious physiological analogy!)

Following this "sperm and egg intercourse," our happy couple takes the next several days to honeymoon their way toward the uterus, where—now joined as one zygote—they set up house together in the endometrium, where the developing embryo will live until time to be born. You couldn't dream this stuff up. There's almost no end to what we take for granted about sex.

From beginning to end, "Ripley's Believe It or Not" doesn't hold a candle to all the incredible aspects of human mating and sexual reproduction.

Sex Behind the Scenes

So far, we've mostly been describing the "delivery systems" for eggs and sperm. If you think those systems are mind-boggling, step down to a deeper level and you'll be all the more gobsmacked by what is actually being delivered and how it comes into being. Take, for example, the metabolic process of *spermatogenesis*—a process which begins for the male at puberty and continually produces millions of sperm each day to make up for the constant loss of those little fellas, who don't like hanging around for long. (Something of a use-it-or-lose-it syndrome.)

Ever hear someone say, "He's half the man he used to be"? If that's true, he just may be the sexiest man around! Let me explain. Sex is all about division, more specifically about dividing in half. This precise percentage should not be taken for granted: It's by *half*, exactly *half*, not thirds, fourths, or fifths. In fact, in the process of *spermatogenesis*, two of the three distinct stages involve dividing the cells in half. (If you don't have a science background, bear with me for a couple of pages of somewhat technical discussion, and we'll be off and running again. Permission granted for quick skimming.)

In the first stage, original **spermatogonia cells** (note: each with 46 chromosomes) line the walls of the seminiferous tubules in the testes. On command from the male hormones, each of the spermatogonia makes a duplicate of itself by a process of exact-copy *mitosis*. What results are **primary spermatocytes**, each of which, like the original cells, has 46 chromosomes.

Stage two is where the real magic begins. By a process of what might be called "divide and conquer" *meiosis*, each primary spermatocyte produces two **secondary spermatocytes**, each of which has (can you guess?) 23 chromosomes—exactly half that of the primary spermatocyte. Repeating that same process, each of the two secondary spermatocytes in turn divides in half, producing a total of four **spermatids**, each of which also has those crucial 23 chromosomes. And not just any random 23 chromosomes, but *one out of each pair* of the non-sex chromosomes.

Spermatids might well be called sperma-*kids*, since they must "grow up" to have a head, a middle piece, and that signature tail (the flagellum) for swimming. Once mature, the spermatids become full-fledged spermatozoa, which we've nicknamed "sperm" (and also refer to as "gametes," the male sex cells which eventually combine with female gametes, or eggs). All that remains is to hear the track announcer say: "Gentlemen, start your engines!"

Pop quiz: How many chromosomes does each sperm have? (You'll quickly see why that's important.)

While all of this clever *spermatogenesis* is going on in the male, the female is going through her own ingenuous process of *oogenesis* (think eggs). The weird thing is that human females are born with a lifetime supply of "eggs-in-waiting" known as **oocytes**. All the "eggs-in-waiting" are right there in the basket *from the moment our female was herself a developing fetus*!

Why eggs *in waiting*? Because the female's eggs aren't fully developed right away. In fact, a woman is born with thousands of oocytes of which only about 500 eventually develop as full-blown eggs. As with the production of sperm, but far more complicated, there are intermediate stages to be gone through. (Here comes more strange terminology, so do your best to stay with me!)

In stage one, through a process of exact-copy mitosis, fetal cells known as **oogonia** turn into cells called **primary oocytes**, each of which has our now-familiar 46 chromosomes. (Again, this is happening all the way back at the fetal stage before the mother-to-be emerges from her own mother's womb! Too weird? Think, "Back to the Future.") But just as the primary oocytes are beginning the process of meiosis, suddenly, everything shuts down until the female enters puberty, as if the oocytes are frozen in time.

At puberty, as everyone knows, the hormones kick in and begin to kick out eggs. Well, sort of. Puberty, of course, is when the menstrual cycle begins, as well as ovulation whereby an egg is released from one of the ovaries midway through each menstrual cycle. But for that to happen, the primary oocytes must resume the process of meiosis that has been waiting on hold all this time—making sure that one primary oocyte re-enters meiosis every 28 days.

That need for resumption of meiosis brings us to the second stage, at which point each primary oocyte does what we expect: it divides into two cells. What we *don't* expect is that these two new cells aren't equal. One cell, a **secondary oocyte** (aka a daughter cell), is larger because it gets most of the cytoplasm with it nutrients, organelles, and mitochondria, *etc.*—all of which it will pass on to the fully-developed egg. The other, tiny little cell is called the **first polar body** (don't ask why!). Relative size notwithstanding, if you had to guess, how many chromosomes does each cell have? You're right, 23—half the original chromosomes (and therefore half of the genetic information of the parent cell), which is what we always get from the "halving" process of meiosis.

Stay with me, we're almost there. The third and final stage involves a second round of meiosis, much like we saw in *spermatogenesis*. Speaking of which, the winning sperm is like an alarm clock going off to remind the secondary oocyte that it's time to hit the throttles on the third stage, which takes place after fertilization in the fallopian tube. Here, the secondary oocyte (can you predict it?) divides itself into two, producing an Extra Large, Grade A **egg**, and its tiny counterpart, a **second polar body**. Meanwhile, the first polar body divides into (this you likely *wouldn't* predict) two new **polar bodies**!

Doing the math, we end up with the one lovely egg we've excitedly been waiting for, and three polar bodies, which after these fascinating cameo appearances simply wither away, thank you very much. In terms of the number of chromosomes, our lovely egg has (you've got it!) precisely 23—which, of course, is exactly what we need from the female gamete (the egg) to match the male gamete (the winning sperm) with its own 23 chromosomes. A match made in heaven if ever there were one.

By the time the egg and the sperm get their zygote thing together, we're back full-circle to the full complement of 46 chromosomes. And the beat goes on...generation after generation.

Ain't sex fun? In fact, there's yet another level we really ought to dig down into, and I'd be sorely tempted to keep going but for my fear that you might put the book down at this point and never get to the heart of the matter in the next chapter. Any readers trained in biology will know what I'm referring to and will be familiar with the finer points of mitosis and meiosis. It's head-spinning stuff! If you are not a science specialist but want to dig just a bit deeper, check out the footnote. (Even there, I promise to provide nothing more than the headlines.)[°] If you want to know more, the books are all there. (Or, if you insist, Google it!)

° Mitosis is broken down into four phases: the Prophase, the Metaphase, the Anaphase, and the Telophase, together with a process known as cytokinesis. By contrast, meiosis is dancing the two-step: Meiosis I and Meiosis II. In **Meiosis I**, Prophase I involves *synapsis* and *crossing over*, which is followed by Metaphase I, Anaphase I, and Telophase I. (The profound mystery of the *crossing over* of genes deserves an entire chapter! In fact, it deserves the spotlight, since nobody, but nobody can explain its origin with a straight face. And there's so much more to say about *cytokinesis*, the incredibly fascinating change in cells...!)

In **Meiosis II** (far more complicated, I assure you, than merely giving it these labels) there's Prophase II, Metaphase II, Anaphase II, and Telophase II. Nice parallels with

You Ask "Y"?

Another important aspect of sexuality we too often take for granted is *gender*. Common to all birds, insects, and other animals, gender—like meiotic reproduction itself—is a crucial hurdle for non-directed evolution to cross.

Simply as a matter of description, we could begin where George Williams does when he says that "the essential difference between the sexes is that females produce large immobile gametes [eggs] and males produce small mobile ones [sperm]." So here's the question: What accounts for this fundamental difference, especially given the fact that "in nonreproductive aspects of life, a male and a female of the same population usually have similar ecological relations: roughly the same habitats, diets, diseases, and so on"?[103]

So far, we've spoken of *male* and *female* as if they just appeared from out of nowhere via the stork or from the cabbage patch. They didn't. Not even Mendel, the pea man, had that one figured out, nor Thomas Morgan, for all his important genetic research with fruit flies. What we do know is that human chromosomes occur in structurally identical pairs...except for the last two, the sex chromosomes, which have been coded as the "X chromosome" and the "Y chromosome." In humans, the female has two X chromosomes, while the male has one X and one Y. (How did *that* happen!)

So, if we go back to the half-and-half sharing of chromosomes between the male gamete (that lucky winning sperm) and the female gamete (the egg), among the 23 chromosomes that each party brought to the relationship there would have been an X chromosome from the egg

---

mitosis, but they're as different as chalk and cheese. The primary difference is that meiosis has that unique, defining characteristic of dividing its chromosomes precisely in half (producing *haploids*, with only one copy of each type of chromosome) which then recombines into the full contingent of chromosomes (back to being *diploids*, with two copies of each type of chromosome).

and, from the sperm, either an X or a Y. Then comes the "chrome mixing bowl," where the chromosomes are randomly blended together to produce that next, wholly unique individual soon to be earth's latest inhabitant. But not just any individual. A *male* individual or a *female* individual! The burning question is not just the excited parents' pre-sonogram curiosity as to whether it's a boy or a girl, but *why gender*? (Indeed, *how* gender?)

Mystery abounds, but this much we know: The female could contribute an X chromosome to make either a son or a daughter; and the male could contribute an X chromosome to make a daughter; but solely the male could contribute his Y chromosome to make a son.

Now don't jump to any huge philosophical conclusions on this one, but the male's unique Y chromosome is smaller than all the other chromosomes (if size matters to you). Even so, it packs a wallop! The Y chromosome is macho to the max, explaining why, from the word *go*, little boys are trouble with a capital "T", which stands for *testosterone*! Because of the male gene in the Y chromosome, the little guys begin to develop their testes six weeks from conception (producing all that testosterone) and, before you know it, come out of their mothers kicking a football!

And Your Point...?

What I've tried to do in this chapter is to convey, if far too superficially, at least some sense of the profound complexity and interactive necessity that underlies sexual activity and reproduction. However, once again I don't want you to be misled about where I'm headed. I'm not building an argument here for so-called Intelligent Design based on "irreducible complexity." That's somebody else's debate. Rather, in furtherance of the narrow purpose of this book, I'm trying to help us focus tightly on the single, crucial issue: *How could the uniqueness of male/female (meiotic) sexual reproduction ever have been derived from the clearly distinctive process of non-gendered asexual (mitotic) duplication, a supposed predecessor process which is the widely-assumed, clearly indispensable, "common ancestor" linchpin in the Grand Theory of Darwinian evolution?*

Without ever directly tackling the issue himself, time and again in his writings Charles Darwin sets up the crucial problem we're addressing in this book, as seen in this passage:

> I believe that animals are descended from at most only four or five progenitors, and plants from an equal or lesser number.... Analogy would lead me one step farther, namely, to the belief that all animals and plants are descended from some one prototype.... With all organic beings excepting perhaps some of the very lowest, sexual production seems to be essentially similar. With all, as far as is at present known, the germinal vesicle is the same; so that all organisms start from a common origin.... Therefore, on the principle of natural selection with divergence of character, it does not seem incredible that, from such low and intermediate form, both animals and plants may have been developed; and, if we admit this, we must likewise admit that all the organic beings which have ever lived on this earth may be descended from some one primordial form.[104]

I fully appreciate that no evolutionist would argue that somewhere along the eons of evolutionary time there was suddenly a giant leap from some single-cell protozoa (or protist-like ancestor) to anything like the unique human reproduction such as I've described in these opening paragraphs. Quite to the contrary, the evolution hypothesis is that the transition happened by an imperceptible gradualism over millions, perhaps billions, of years. And, furthermore, that the very first sexually-reproducing organisms would have been primitive life forms which, over still more blurring eons of time, evolved into the complex sexual beings we humans are today.

This explanation is confirmed by Carl Zimmer and Douglas Emlen in their encyclopedic text, *Evolution—Making Sense of Life*: "Nor do evolutionary biologists claim that natural selection depends on many mutations arising all at once. Instead, mutations can arise and spread one at a time."[105]

Given the functional uniqueness of sexual reproduction *at even the most primitive level*, what we will see over and over throughout this book is that such an assumed gradual process could not, in actual scientific fact, have happened.

At the very least, Richard Dawkins himself recognizes what it takes to move from simple to complex. "Whether the replicators that are selectively eliminated are genes or species, a simple evolutionary change requires only a few replicator substitutions. A large number of replicator substitutions, however, are needed for the evolution of complex adaptations."[106] Surely, sexual reproduction is a prime example of a complex adaptation for which a large number of replicator substitutions would be required.

The complexities of human reproduction—particularly at the cellular and molecular level—are simply a reminder of the complexity that lies at the base of all sexual reproduction no matter how primal. Which is not to say that the process of mitosis in asexual replication has no elegant complexity of its own, begging the question as to how such wondrous replication itself might have arisen over evolutionary time. It's just that the particular, distinct complexity of asexual (mitotic) replication is incapable of introducing the particular, unique complexity required for male/female (meiotic) sexual reproduction.

On the road to male/female meiosis, natural selection can't *select* from mitosis the required gender contributions and sexual DNA which mitosis never had to begin with. Nor can any of the more-unusual "para-sexual" permutations of asexual replication give a leg up to the altogether different blending of male and female gametes in male/female meiosis.

It's important to emphasize that asexual replication and sexual reproduction do not differ simply by *degree*, but *in kind*. Whereas the former might arguably be bridged over time, the latter could not possibly have been bridged given all the time in the world. The difference between *degree* and *kind* is all the difference between exact-copy *mitosis* and the unique mixing of male/female genes in *meiotic*, sexual reproduction. Those two processes are not simply different degrees of the same action, but two completely different ways of getting from one generation to the next.

Revisiting the "Generation One" Problem

In fact, that very distinction is why we can't possibly get from one process (mitosis) to the other (meiosis) in evolution theory. *To get from asexual replication to Generation One of sexual reproduction, every single detail of even the simplest sexual reproduction would have to be firmly in place.*

I'm not unaware that this "Generation One" problem, as I call it, is a huge sticking point for evolutionists, and probably the hardest sell of anything in this book. Virtually by definition, anything other than gradual change over time is an anathema for evolutionists, challenging as it does the most fundamental premise of evolution. If you've already bought heavily into evolution theory, it makes perfect sense that anything like a "Generation One" rule must surely be wrong. By evolution theory, absolutely nothing evolves in, or depends on, a single generation alone. Doesn't everybody know that?

Absolutely. Everybody knows that. Even I know that. But that's exactly my point. If there's something that couldn't possibly happen gradually over multiple generations, then we've got a serious problem with evolution theory. So, is there anything that couldn't possibly happen except in a single, fully-developed generation? (Or, if you prefer, in anything less than a *fully-evolved*, completed process?) Through a series of hypotheticals, I'm going to let you answer that question for yourself, starting with some rather quirky scenarios:

1. Suppose your mother and father hadn't...you know...gotten together. Would you be here as their child?
2. If someone were to say to you, "I know this sounds incredible, but, truth is, you were produced over three generations of your ancestors, not in just one generation," what would you say to them?
3. Suppose you go back in time to when your mother was herself a developing fetus and your father was still in his mother's womb. What are the odds that they could have given you birth in their nascent states of existence?
4. Suppose your father was 18 when your mother was still a developing fetus. Any chance of your being conceived at that point in time?

To bring the questions closer to evolutionary biology,

1. Suppose two asexual organisms somehow evolved to be capable of mingling together their separate, distinct genes. Does that give us male/female meiosis?
2. Suppose an asexual organism replicating by mitosis has a series of mutations that, over many generations, evolves against all odds into an organism having female features. Does that get us to meiosis?
3. Suppose another asexual organism simultaneously evolves into an organism having male features. Does that get us to meiosis?
4. Suppose those two male and female organisms, having evolved some primitive instinct to mate, have a different number or kind of chromosomes. Does that get us to meiosis?

5. Suppose that each organism evolves to have the same number of (matching) chromosomes. Does that get us to meiosis without each of them also evolving the ability to reduce their chromosomes precisely by half, and mingle their genes through crossing over and recombination?
6. If any one of those factors is missing before the first-ever instance of male/female meiosis, could there possibly be a second generation of male/female meiosis of any kind?

Do you still not see the "Generation One" problem?

Before offering a tentative but strangely inconsistent way to get around the problem,[P] Adam Wilkins and Robin Holliday highlight the "Generation One" problem (with my emphasis):

> *While meiosis almost certainly evolved from mitosis, it has not one but four novel steps:* the pairing of homologous chromosomes, the occurrence of extensive recombination between nonsister chromatids during pairing, the suppression of sister-chromatid separation during the first meiotic division, and the absence of chromosome replication during the second meiotic division. *This complexity presents a challenge to any Darwinian explanation of meiotic origins. While the simultaneous creation of these new features in one step seems impossible, their step-by-step acquisition via selection of separate mutations seems highly problematic, given that the entire sequence is required for reliable production of haploid chromosome sets.* Both Maynard Smith (1978) and Hamilton

[P] In something of a reversal from their opening acknowledgment, Wilkins and Holliday conclude their study, saying, "We have argued that the origins of meiosis from mitosis initially involved only one new step, namely homolog synapsis. Two of the other unusual features of meiosis are prefigured in mitosis and would have been brought into play as consequences of the existing regulatory features of mitosis while the remaining one (extensive recombination) could have evolved later." *Prefigured* is hardly the same as *precursor*. And how can one of the four *absolutely indispensable* components have evolved *later*...?

> (1999) regarded the origins of meiosis as one of the most difficult evolutionary problems.[107]

(W. J. Hamilton's candid confession is worth a brief pause: "*...if there is one event in the whole evolutionary sequence at which my own mind lets my awe still overcome my instinct to analyse, and where I might concede that there may be a difficulty in seeing a Darwinian gradualism hold sway throughout almost all, it is this event—the initiation of meiosis.*"[108])

You say that various life cycles prior to male/female meiosis might have provided the building blocks leading to that meiosis? Unless *all necessary functions* could have been brought into play simultaneously, you'll never get to your destination. Imagine a pile of steel girders, gigantic bolts and nuts, miles of thick cable, and countless sacks of concrete lined up on one shore of San Francisco Bay. Does that give us the Golden Gate Bridge? Suppose we're only able to construct the bridge a quarter of the way across for lack of engineering information on how to secure the upright supports in the deepest part of the bay? Are you prepared to drive onto the completed portion of the bridge, hoping your car can make the unthinkable leap all the way to the other side?

Evolutionist Mark Ridley confirms the necessity for the full "package":

> Meiosis, gene transfer and gender are all connected. They form the basic set-up of Mendelian reproduction.
>
> Bacteria and other simple life forms mainly clone themselves, though some of them do occasionally use certain kinds of mock-sexual gene-shuffling. But they all lack the special cell division of meiosis, and the kind of systematic, fair-exchange, gene-shuffling, gendered sex that we and other complex life forms use.[109]

A Journey Back in Time

That last line, "we and other complex life forms," is important. There is no contention in this chapter that human reproduction is to be regarded as archetypical of all sexual reproduction. It is merely a particular case—and the most familiar—illustrating the complicated inner workings of meiosis in typical sexual reproduction, and graphically demonstrating the incalculably-complex requirements for even the simplest male/female sexual reproduction possible.

More important still, the complexity of meiosis in human reproduction illustrates how the first-ever occurrence of male/female meiosis would have been so radically, distinctively, uniquely different from the supposed precursor process of mitosis, including any variation on theme you might wish to imagine. Can one point to similar features between mitosis and meiosis? Sure. In fact, doesn't mitosis itself occur in the process of meiosis? Absolutely. But if the transition from mitosis to meiosis were so easily explained by those similarities, we wouldn't be having study after study defying the best of minds trying to figure out how the one process could possibly have evolved into the other.

The challenge, therefore, is to take what we know about the microbiotic intricacies of human reproduction and re-play the supposed tape of evolutionary history all the way back through primates and pre-primates (whatever they might have been), and down the supposed tree of evolution—branch after branch, through the unique sexual reproduction of birds, fish, and plants, and on down through the earliest sexually-reproducing life forms we can possibly imagine until we stand right next to the first-ever male and first-ever female of any type, at the first-ever moment when two living organisms of different sexes joined together in the first-ever occurrence of male/female meiosis to produce the first-ever gendered offspring that in turn produced the next generation of sexually-reproducing organisms.

If we found ourselves standing next to that very first gendered pair on our right, what would we see if we looked back to our left? If it's some asexual organism with not a hint of maleness or femaleness, nor the slightest whiff of sexual DNA, then where did our first-ever male/female meiosis on the right come from?

If someone suggests that what we see on our immediate left is a partially-developed male or partially-developed female, or partially-developed process of meiosis, how possibly could any of those partial steps have survived from one generation to the next, especially given the cloning nature of asexual mitosis?

If the proposal is that there might have been a diversity of asexual life cycles enhancing and embellishing common, ordinary mitosis which might have been the precursors to male/female meiotic reproduction, how many different assumptions would be required to come up with both the first-ever male and female independently (or more demanding yet, in tandem); having novel genetic information never known in asexual creatures; with compatible chromosomes both in number and type; together with a complicated process of creating haploid cells from diploid cells (demanded by your basic meiosis); with the never-before-seen ability to have "crossing over"; or developing a revolutionary process of restoring haploid cells to diploid cells in a unique offspring that is fully male or fully female, having the same full complement of chromosomes as their parent organisms?

Might you suggest any other scenario? Are you confident that it eliminates all the daunting problems, or is it just another shot in the dark? In the pages ahead, we will explore most if not all of the many theories that have been proposed to explain the evolution of male/female meiosis and sexual reproduction from what is assumed to be its precursor process of asexual mitosis. You can judge for yourself, but I think you'll find each of those theories falling woefully short.

At least tentative confirmation of that conclusion comes in the consensus of the scientific community itself that, despite all the valiant attempts to solve the conundrum, there is no viable evolutionary explanation for the existence of male/female meiotic sexual reproduction. Virtually all of the putative models currently being suggested end up acknowledging the obvious: "The origins of meiosis in early eukaryotic history have never been satisfactorily explained."[110] And again from current literature, "The very existence and persistence of sexual reproduction remains an evolutionary puzzle."[111]

As we've already discussed at length, there is admirable honesty in those candid admissions, but also futility in the misplaced hope that one day the puzzle will be solved. To repeat a familiar, but important, line: *Given the very nature of gendered sexual reproduction*, natural selection could not possibly have provided, either gradually or abruptly, an evolutionary bridge between any form whatsoever of non-sexual replication and the unique male/female meiosis required for sexual reproduction—without which bridge Darwin's Grand Theory could not possibly survive.

Recap

- *Asexual replication* is a process of duplication known as *mitosis*, which, apart from any (usually unhelpful) mutations, results in cloning. (Think "copy machine.")

- *Sexual reproduction* is a process known as *meiosis* wherein an organism's chromosomes divide in half, then mix their DNA to produce individual offspring different from the two parents and every other organism that has ever existed. (Think "blender.")

- Lacking any sexual DNA, *mitosis* could not have provided either the information or mechanisms required for the radically-different process of *meiosis*.

- The amazing details of human reproduction illustrate the intricacies of *meiosis* in even the most primitive forms of sexual reproduction—a process so radically distinct from asexual *mitosis* that no evolutionary process—whether gradual or abrupt—could make such a dramatic transition.

- Despite no end of theories and speculation, the scientific community admits it is completely baffled regarding the "Queen of evolutionary problems." And that's not simply because they haven't *yet* discovered the bridge between asexual and sexual reproduction, but because no bridge is scientifically or logically possible.

Question: Do you believe that the second-ever generation of sexually-reproducing organisms could have come into existence if the essential components required for the first-ever occurrence of male/female meiosis could not have evolved all at the same time through a gradual process of evolution?

CHAPTER 5

Not Even the Sexiest Asexual Could Ever be Sexual

Amoebas at the start were not complex—
They tore themselves apart and started sex.
—Arthur Guiterman

Presenting the fundamental problem of "the missing sex link"—the crucial developmental gap between asexual replication and sexual reproduction—and highlighting the necessity of making that transition in a single, historic generation rather than by progressive, slight modifications from the distinctiveness of mitosis to the uniqueness of meiosis.

Let's be honest: single-celled amoebas aren't very sexy. Perhaps they have a certain winsomeness about them, but not a bit of what we ordinarily think of as sex. There's not a hint of masculinity or femininity, not an ounce of seductive instinct. Amoebas replicate themselves asexually by *mitosis*, dully producing exact copies of themselves. Boring, really. So, the question is: How did evolution ever get to the point of providing sexier life forms—from those "loser" amoebas to exciting creatures that mate and reproduce sexually—which is to say by a process of *meiosis*?

Undoubtedly, American Poet Arthur Guiterman's little ditty in the chapter quote above was intended merely to be clever and catchy, but it actually raises a vital question: How possibly could amoebas have "torn themselves apart" in the way that would have been required for reproducing in a manner never before known in evolutionary history? As you can well imagine, this is the heady stuff of evolutionary research.

When it comes to sex, it's difficult to keep things simple. Sex is complicated! For instance, even the story of the lowly amoeba is not quite as simple as the opening paragraphs might lead you to believe. There's a certain matter of classification to consider. Depending on how you classify various micro-organisms, you might just find some incredibly sexy "amoebas" alongside, say, the notoriously chaste *Amoeba proteus*—that venerable star of stage, screen, and standard biology texts. At least one study posits that we may have to re-think what kind of "amoebas" we're talking about rather than indiscriminately tarring them all with the same asexual brush.[112] Apologies all around to any amoebas we may have offended. However, we do have at least a flimsy defense. The smaller the organism, the more difficult it is to be a sexual voyeur (all in the legitimate service of science, of course!).[q]

Should anyone be so indelicate as to raise the potential amoeba sex scandal, it's easy enough simply to say that in the present discussion we're talking only about those amoebas that are unquestionably asexual.[r] For our purposes, it doesn't really matter which particular asexual prototype it might have been in the putative chain of Darwinian evolution. By Darwin's Grand Theory (hypothesizing the common descent of all living forms from simpler life forms), it would have been essential for

[q] There is, further, the not-so-small matter of defining just what constitutes "sex." Does meiosis alone count as sex (going "halvsies" and coming up with reproductive gametes), or must we also observe karyogamy (recombination whereby the two become one). Would it be "sex" if we could observe some form of karyogamy, but no meiosis leading to it? (In such a case, odds are that reproduction would be happening in some fashion other than normal sex, but one thing remains certain: we're still not talking about simple, asexual duplication.)

[r] If it's true that there are sexually-reproducing amoebae, the central question of this book remains: How did those amoebae ever evolve from the original asexual progenitor?

a clearly asexual prototype *of some kind* to be the radical change agent for all its sexually-reproducing descendants.

Call it a microbe, if you prefer; and think of evolution as "microbe-to-man," if you wish to be precise. But since the word "amoeba" means *change*, it's only fitting in this chapter that we should stick with that humble creature as our asexual prototype, even if in actual fact it might have been some other asexual life form that made the great leap to sex. In the spirit of the rollicking stage play "No Sex Please: We're British!", our working assumption will be, "No sex please: we're amoebas!"

Handling "Sex" Carefully

Before proceeding further, we need to clear up some definitions. In common parlance, the word "sex" might refer to 1) simply gender—as in male and female; or 2) *having* sex—as in sexual intercourse; or 3) *meiosis*, which is a particular process of diversifying the genome wherein there is a complicated exchange of genetic material between chromosomes involving what is known as "crossing-over" and "recombination." For the moment, let's focus on this last meaning: *meiosis*.

Put technically, *mitosis* starts and ends with diploid cells, meaning two copies of each type of chromosome.[s] By contrast, *meiosis* starts with diploid cells then reduces down to haploid cells, meaning one copy of each type of chromosome (i.e., gametes, or sex cells). This fifty percent reduction of cells prepares the gametes for fusion, thereby enabling the process of meiosis which involves the crossing over of shared genes followed by recombination, the result of which is a return to the diploid state.[113] Put far more simply, mitosis is a cookie cutter turning out predictable sameness, while meiosis is a blender from which highly varied offspring emerge.

---

[s] In some cases, there may be haploid cells (as in the gametophyte generation of mosses, ferns, and fern allies), but the point remains: mitosis and cytokinesis can only produce two copies of the starting cell.

Clearly, meiosis can occur without either "sexual intercourse" or, strictly speaking, the need for anatomically-recognizable males and females. However, since that is not the usual case, the discussion which follows generally assumes the more typical male/female sexual mating and reproduction (or, in some cases, external fertilization). Talk all you want about interesting life forms associated with asexual replication, or the diversity of life cycles. Nature is full of wonder and mystery. But there's no mystery that, according to evolution theory, at some historical point in time there had to be a first time for the kind of sexual reproduction so ubiquitous throughout Nature.

For purposes of this book, the sole issue regarding the "origin of sex" is this: How did male/female sexual mating and reproduction arise in the very first organisms that started procreating by male/female reproduction?

The problem for evolutionists is explaining how, by any means over any amount of time, this particular kind of sexual reproduction possibly could have evolved from any form of asexual replication one might wish to imagine. If all you have to work with is mitosis, there are three huge hurdles to cross: 1) There's no DNA information for sex; 2) there's no male/female gender distinction; and 3) there's no process of male/female meiosis.

When it comes to mitosis and meiosis, you can't just mix and match. Graham Bell puts it simply: "the absence of any sexual process leads to the replication of identical progeny from a single ancestor."[114] With mitosis, (apart from the odd, rare mutation) what you see is what you get. With meiosis, what you get is anybody's guess.[t] Maybe some family resemblance, but you can't count on it.

George Williams highlights the differences between the two crucial processes by focusing on generalizations which can be applied to their offspring: "Asexual progenies are mitotically standardized, and sexual

[t] When applied to a few specific gene loci (or alleles), Mendelian laws enable us to go beyond guessing.

ones meiotically diversified."[115] Generally speaking, asexual offspring are of initially large size, not small; are produced continuously, not seasonally limited; develop close to the parent rather than being widely dispersed; develop immediately, not dormantly; develop directly to the adult stage, rather than through a series of diverse embryos and larvae; have a genotype predictable from the parent as opposed to unpredictable genotypes; and have a low mortality rate as compared to the higher mortality rate of offspring produced meiotically.[116]

Of one thing we can be sure: barring mutations, mitosis can never be more than a glorified copy machine. By contrast, the very nature of male/female meiosis enables it to produce sexually-diverse progeny.

Another thing to keep in mind: We must be careful not to confuse true male/female *meiotic* sexual reproduction with other forms of reproduction involving genetic exchanges, such as the microbial "sex" process whereby genetic material is exchanged through a horizontal gene transfer. In bacteria, this involves swapping two segments of DNA through a process known as "transduction," or, more commonly, through "conjugation" whereby two microbes nudge up to one another, momentarily joining together, or perhaps even suspending a tube (or bridge) between them so that one microbe donates its gene to the other.[u]

Such identical, or nearly identical, "cloning" is fundamentally asexual in nature.[117] In no way are these particular microbes "male" and "female;" the tube is nothing like a penis transferring haploid male gametes intended to merge with a haploid female egg; and—most

---

[u] To avoid any confusion, there is also a different type of "conjugation" that is seen in the single-celled *Spirogyra*, among the green algae, which involves a form of "mating" in which one strand or filament is the donor and the other strand is the recipient. Even if "donor" and "recipient" are suggestive of "male" and "female," *Spirogyra* undergo mitosis, effecting *asexual* replication. (This is not to say that Spirogyra lacks the capacity for sexual reproduction where, for instance, meiosis can occur in harsh conditions. But even there, the question remains: how would the capacity for sexual reproduction have evolved?)

importantly—there is no intermingling of male and female genes capable of producing a new, unique, diploid offspring that is itself either male or female.[118]

And let no one be deceived: No perceived advantage from any variant of asexual replication can be a gradual stepping stone to a radically different form of reproduction requiring numerous sophisticated processes never associated with any version of asexual replication. Taking two steps *away* (from mitosis) doesn't put you two steps on your way *up* (to meiosis). Just because improvements might be made to a Ford doesn't mean that it could ever turn into a Ferrari. No amount of "sexing up" of mitosis will get you anywhere near the meiosis of male/female *sex*.

There are always those who cling to the belief that if *any* changes are possible, then *all* changes are possible. In this particular context, they insist that if we have any examples of organisms that mimic even a miniscule part of what's involved in male/female meiosis and sexual reproduction, then surely being part way along the process opens the door to the entire process. By some irony, it's the same "if-any-changes-then-all-changes" fallacy that is at the heart of Darwinism. Observing change within established species, Darwin made the extraordinary leap of concluding that all species resulted from that self-same process of change. To this day, evolutionists are so steeped in Darwin's grossly overextended extrapolation that there is no part of the Evolution Story immune from its blinding assumption.

We'll speak more of this elsewhere, but there is even less reason to apply the "if-any-changes-then-all-changes" extrapolation to the origin of sex. Unlike the incremental gradualism required for Darwin's Grand Theory, the gap between mitotic, asexual replication (no matter how embellished or oddly diverse) and the first-ever male/female meiotic sexual reproduction could not possibly be bridged in any gradual, step-by-step fashion from Point A to Point B.

To understand why, merely break it into its component parts. What must you absolutely have to have for the first-ever male/female meiotic sexual reproduction? 1) A male capable of moving from diploid to haploid, producing male gametes. 2) A female compatible in chromosome count, mating capability, and reproductive processes. 3) Never-before-seen meiosis, crossing over, and recombination. Do you think that any one of those component parts (or sub-parts) alone could have produced the male/female offspring required for any succeeding generation?

No one can leap the Grand Canyon step by step!

For genuine male/female meiotic reproduction, anything half-way is a tease. To have real, genuine sex, you've got to go all the way.

Loose Use of "Sex"

It's not helpful when writers use the term "sex" loosely, as when Mark Ridley (consistent with his refreshingly breezy style of writing) says, "Bacteria do use sex, or mechanisms rather like sex, as we have just seen; but bacteria are mainly clonal."[119] Since Ridley's immediate follow-up is: "When a bacterium divides it usually passes on copies of the same DNA as it inherited," why talk about "bacterial sex"? Mechanisms of DNA transfer in bacteria may have some sex-like similarities, but they don't come close to meiotically mingling together distinctive DNA in the same way as two separate genders in sexually-reproducing organisms. Despite their shared similarities, there can only be confusion in calling a daffodil a rose. Why, then, confuse any process of non-sexual cloning with genuine sex? As evolutionist Nick Lane reminds us, "no bacteria have true sex."[120]

Loose use of the term *sex* is one of the primary problems with the analysis offered by Lynn Margulis and Dorion Sagan in their book, *Origins of Sex*. We're immediately put on notice that they have a rather expansive view of "sex" when they say in the Preface to the second printing that "Some aspect of sex in one of its guises is probably as old as life itself."[121] I suppose if you consider any genetic mingling of any kind to be "sex," then that conclusion reasonably follows. But talking about that kind of "sex" in the context of trying to explain the origin of male/female meiotic "sex" simply muddies the water, especially since no causal connection can be made between the two distinct processes.

Trying to nail down what kind of sex the authors are talking about is like playing a game of hide-and-seek, typified by this discussion:

> Sex is a genetic mixing process that has nothing necessarily to do with reproduction as we know it in mammals. Throughout evolutionary history, a great many organisms offered and exchanged genes sexually without that sex ever leading to the cell or organism copying known as reproduction. Although additional living beings are often reproduced by a combination of genes from more than a single parent, sex in most organisms is still divorced from growth and reproduction, which are accomplished by nonsexual means.[122]

Sex, untied to reproduction? While I appreciate the "sex-like" intricacies of fusion, genetic exchange, and horizontal transfers, that's not the kind of clearly reproductive sex we're talking about in this book. Scholarly discussions about the "origins of sex" invariably invite confusion by including references to *reproductive* sex alongside discussion of what is said to be sex unassociated with reproduction. Consider, for example, this reference to both asexual bacteria and sexual protists:

> Sex first appeared in bacteria. Later, in larger, more complex microbes called "protists," a new and different kind of sex evolved.[123]

Are bacteria *sexual*? Do bacteria *reproduce* sexually? If not, why refer to "bacterial sex" or "bacterial sexuality"? The answer comes in their definition of sex: "By *sex* we mean...the complex set of phenomena that produces a genetically new individual, an individual that contains genes (genetic material, DNA) from more than a single source."[124] By that definition of sex, *asexual beings* can be *sexual beings*!

I appreciate that one can define "sex" any way one wishes in an effort to understand sexual mysteries large and small, but I confess I wonder if evolutionist authors prefer an expanded definition of "sex" whenever possible in order to give them some wiggle room in the truly daunting task of explaining the origin of male/female meiotic "sex." If you can think generically of "sex" as a *continuum* of various kinds of reproduction, then there's less pressure to explain a unique process like meiosis that is dramatically different and functionally distinct.

At least the crucial distinction finally emerges in the following passage from Margulis and Sagan:

> Bacterial sexuality is very different from the meiotic sex of protists, fungi, plants, and animals, and it evolved far earlier. Meiosis, or cell division resulting in reduction in the number of chromosomes, and subsequent fertilization, or reunion of cells to reestablish the original chromosomal number, first occurred in protists.[125]

Consistent with that distinction, it should be noted that the process of meiosis associated with sexual reproduction may be seen in simple life forms (like *fungi* and *protists*, including algae) that don't plainly exhibit "male" and "female" characteristics. For instance, there are the more animal-like protists (i.e., *protozoa*) which lack typical "male" and "female" morphological distinctions (i.e., body shape) and therefore are simply referred to as (+) and (-) forms. But, importantly, the process is one of meiosis, not mitosis.[v]

---

[v] In plant-like protists such as Spirogyra, a similar distinction exists between donor and recipient filaments which join in conjugation to form zygotes, which in turn undergo meiosis.

If in various organisms there need not be "sex," strictly speaking, for reproduction, there remains a crucial difference between asexual replication[w] and variety-producing forms of genetically-mixed (sexual) reproduction.[126]

That Peculiar Parthenogenesis

Before you raise your hand to ask, yes, you're right. Whiptail lizards in the southwestern United States reproduce by a process of meiosis involving a recombination of chromosomes (whereupon the egg begins to develop into a lizard embryo) but do not require DNA from sperm in order to reproduce.[127] Parthenogenesis is relatively common in reptiles. Anomalous though it may be not to have male input, we're still faced with the radical difference between mitosis and meiosis, and all the other features of sexual reproduction including eggs and fertilization never known in a completely asexual world. So, the fundamental conundrum remains, not to mention that evolutionists do not point to the whiptail lizard as the common progenitor of all things sexual.

Parthenogenesis may also be present in organisms whereby in a single species (like aphids, hydras, and sponges) there can be both asexual and sexual reproduction, usually in cycles (begging the question of how the *sexual* cycle ever came to be).

In a variation on theme, Aristotle noted another kind of parthenogenesis (as with bees), whereby the development of embryos occurs without immediate fertilization of the egg cells by freshly-donated sperm. That's code for "No males need apply," thank you very much. (The poor ol' drones that provide the semen for the queen bee are virtually emasculated, given that they never have a full chromosome count.)

---

[w] Qualified, perhaps, by such phenomena as antibiotic resistance in a small percentage of a bacterial population.

No mystery then that these anomalous female-oriented reproductive processes wear the label *parthenogenesis*, which is Greek for *virgin birth*. (How delightfully ironic it would be if any form of reproduction were based upon evolutionary belief in a virgin birth!) But nothing here answers our questions about the origin of sex. We've still got to explain how evolution, acting blindly, moved from genderless asexual beings to any form whatsoever of *female gender*, together with her reproductive *eggs*. Nor is there any ground-swell of support from evolutionists claiming parthenogenesis as the portal through which sexual reproduction emerged. (Ever so curiously, parthenogenesis is frequently classified as *asexual* replication…females, eggs, and all!)

AC/DC Swingers

It's almost impossible to talk about asexual replication versus sexual reproduction without recognizing that there are strange, outlier creatures that can "swing" either way. Consider, for example, the humble yeast which, when food is plentiful, reproduce asexually, via either budding or fission; but, when food is scarce, reproduce by a process of meiosis, forming haploid spores that eventually fuse with other spores to form a diploid zygote. Sounds like sex, doesn't it? To the extent that it *is* sex, the question remains: How did yeast ever acquire its ability to reproduce by meiosis? *Description is not explanation.*

The same question dogs the species of freshwater rotifer known as *Brachionus calyciflorus*, which "reproduces asexually at low densities but sexually in crowded environments."[128] See if you can spot the problem in the following sentence: "When bunched together, the rotifers release a chemical cue that stimulates some females to produce haploid eggs that either develop into males or are fertilized to become diploid females."

Catch it? There is the assumed existence of *females* and *males*, once again begging the nagging question regarding the origin of gender.

More Weird Rotifers

There are rotifers, and then there are rotifers! The latest darlings of evolution theory seem to be the 400 species of *bdelloid rotifers*, which reproduce asexually through a unique process involving dehydration and rehydration (some term it *anhydrobiosis*), during which process they sometimes ingest foreign DNA. Since their closest relatives reproduce sexually, evolutionists have theorized that the *bdelloid rotifers* abandoned sexual reproduction 100 million years ago, returning to a process of asexual reproduction. How very strange, then, that we should read this line from Zimmer and Emlen: "Bdelloid rotifers are all female...and daughters are genetically identical to their mothers." Stranger yet: "They reproduce parthenogenetically, yielding daughter genotypes identical to their mothers, and they completely lack both males and meiosis."[129] Why not simply say they replicate clonally? If it looks like a clone, walks like a clone, and quacks like a clone....

Referring to the ingestion of foreign DNA observed in *bdelloid rotifers*, Mark Welch, an evolutionary biologist at the Marine Biological Laboratory in Woods Hole, Massachusetts, says, "That [particular] transfer may be a surrogate for the exchange that happens during meiosis."[130] Yet, Welch is quick to address the obvious problem being raised in this book: "We still really don't know the answer to this very most basic question. We don't know why sex exists."

As *bdelloid rotifers* remind us, there's all sorts of strange and wonderful "dancing" going on in Nature. But the obvious fact remains: The strange and the wonderful doesn't begin to move us from asexual to sexual. If it did, why would so many scientists and science writers still be admitting they have no idea how male/female meiotic sex came about? Would any evolutionist take the stand and testify that rotifers of any type were the transitional, "common ancestor" intermediary between asexual and sexual?

The hard fact is that meiotic sex is a wholly unique process involving the mixing of genetic information that produces genetic variety in a way that neither asexual fission, fragmentation, conjugation, nor cloning can possibly produce. *Asexual* to *sexual* is not as easy as simply dropping the single letter "a." The two processes could hardly be more different, nor more incapable of a gradual transition from the uniquely asexual to the uniquely sexual.

Mechanism, Mechanism, Mechanism

One of the rare exceptions to the scientific community's general silence about the problem of sex is Sir John Maddox. The title of his book, *What Remains to be Discovered*, ought itself to be a flashing yellow caution light, but listen to what Maddox says candidly about the origin of sexual reproduction:

> The overriding question is when (and then how) sexual reproduction itself evolved. Despite decades of speculation, we do not know. The difficulty is that sexual reproduction creates complexity of the genome and the need for a separate mechanism for producing gametes.
>
> Much more must be learned of the course of evolution before it is known how (rather than why) sexual reproduction evolved, but meanwhile the yeasts provide a clue.[131]

Note carefully Maddox's repeated warning that the issue is not the WHY of sex (for which there's no end of theories, invariably based on its supposed advantages), but the HOW of sex—that is, the indispensable *mechanism* that could possibly have made sex happen. *That* is the single, compelling question prompting this book. And what an irony: Darwin is credited with supplying the *idea* of evolution with an explanatory *mechanism*—natural selection—yet Darwin made no serious attempt to apply his vaunted mechanism to explaining the crucial origin of sex.

Having highlighted the problem, Maddox is less than helpful when he speculates that sex might possibly have resulted from something like the way yeasts transfer genetic material (sometimes haploid, sometimes diploid), saying that "there is a huge gap between this rudimentary form of sexual reproduction and that in the simplest form of animals and plants, yet it is probable that the first led to the second."[132]

Unfortunately, the huge gap to which Maddox refers can't begin to be filled by any analogy to yeast replication. That process of replication is either not anything like meiosis (which wouldn't advance the cause in the slightest) or it is something very much like meiosis—begging the very question of the hour: *How could anything like the haploid process of meiosis have evolved from mitosis?* (Incidentally, calling yeast replication "sexual reproduction" while trying to explain how it might have led to the *origin* of sexual reproduction illustrates once again the careless circular reasoning invariably employed by evolutionists.)

An Eye on Sexual Chemistry

In paying homage to the amoeba and the forgotten millions of microbiotic life forms that too often get ignored because of our love affair with all things big (like elephants, gorillas, and our own human race), botany professor Nicolas Money, pens an elegant line in his book, *The Amoeba in the Room*, saying "the miniature is everything."[133] When it comes to sex, the miniature truly is everything.

Merely consider what's often said of a man and a woman in the context of love and sex: "They had the perfect chemistry!" There's far more to that expression than we might think. Whether we're talking about human beings or seaweeds and mosses, Michael Pitman reminds us that "the sex system is based on an exact programme of chemical switches and signals—mostly in the form of hormones. Not only the correct genetic subroutines and sex organs but the right chemicals in the right sequence are required for satisfactory sex."[134] As with our two lovers who have that head-spinning, heart-thumping "chemistry," the chemistry for sex has to be programmed from the very start. That is, *fully* programmed, not *partially* programmed in fits and starts along some gradual evolutionary path. There may be many circumstances where "half a loaf is better than none;" but for sex ever to have happened, half-way is *no way*!

As everyone knows, Darwin spent many a sleepless night worrying about how something as complex and exquisite as the human eye could evolve gradually over many generations from, say, the light-sensitive pigments of flatworms. Darwin eventually consoled himself with the thought that perhaps those pigments might eventually be coated with a membrane in some intermediate creature, which then developed into a crude lens that ultimately evolved into sophisticated "telescopes" in birds and mammals and, lastly, into the incredible human eye.

In summary, here is Darwin's sales pitch:

> It has been objected that in order to modify the eye and still preserve it as a perfect instrument, many changes would have to be effected simultaneously, which, it is assumed, could not be done through natural selection; but as I have attempted to show in my work on the variation of domestic animals, it is not necessary to suppose that the modifications were all simultaneous, if they were extremely slight and gradual.[135]

Darwin's argument was that, because a little eyesight is better than none at all, each incremental benefit would be retained and built on by natural selection.[136] As wishful as all that is, it's at least more theoretically possible than any attempt to use the same wishful thinking to explain the progression of sex from non-sex. It's not as if you already have "a little sex" that can be selected incrementally as advantageous. You've got *no* sex! (Which is to say—for those with expansive definitions of "sex"—no male/female *meiotic* sex.) Zero times zero is still zero.

At least flatworms had light-sensitive pigments to begin with. And you could always point to some crustaceans which arguably have useful membranes. But with asexual beings, you have nothing whatsoever to work with—not a single sex gene or molecule with potential for onward and upward progression.

This is the precise fallacy behind Darwin's insistence that the complexity of an eye is no big deal for evolution's gradualism and that, if even an organ as complex as an eye could result from evolution, then there is literally nothing that cannot be explained by evolution. What conceivably might work in the development of an eye from primitive light sources (itself an incredible stretch of logic and common sense), cannot conceivably explain how sex evolved *without first having something to evolve from*, beginning with the *information* for sex.

For Information, Dial Zero

Robin Holliday reminds us of the importance of genetic *information* in the perpetuation of the germ line:

> The production of a normal adult organism depends on the formation in the first place of a normal egg, and the normal program for development cannot proceed unless the initial information is correct. Since the genetic material is DNA, it is clear that the information it encodes must be free from errors or defects. The argument can therefore be made that meiosis, recombination, and sexual reproduction are the additional essential mechanisms that ensure the continuity of the DNA in the germ line is maintained.[137]

Not only must the DNA be free from errors or defects, but, rather obviously, the genetic information it contains must already be encoded for meiotic reproduction rather than mitotic replication. You can mix DNA together in meiosis because the necessary information is available from the two contributing chromosomes, but you can't mix together the two radically-distinct processes of meiosis and mitosis because the original process of mitosis could not have been Googled for the DNA information that the previously-non-existent process of meiosis would require in order to get going.

In the process of making another point altogether, Richard Michod confirms the obvious, that

> For any trait to evolve, there must be genes that determine the trait.... So for sex (or asexuality) to evolve, there must be genes that determine whether and how an organism has sex. These are sex genes.[138]

So where did these sex genes come from at the point in evolutionary history when there were only asexual genes available?

In terms of the genetic code, it is a truism that *information* does not originate from any mechanistic process. For meiotic sex ever to have gotten a start, the *information* for meiosis had to come from somewhere... but *where*? Certainly not from asexual organisms which never had the slightest idea, much less *information*, about sex. When Mark Ridley says "The purpose of life is to copy DNA or, to be more exact, information in the form of DNA,"[139] the problem is that you can't copy information that doesn't yet exist.

A failure to recognize this basic problem may explain Ridley's flawed analogy. "The 'sexual' method of reading a book," says Ridley, "would be to buy two copies, rip the pages out, and make a new copy by combining half the pages from one and half from the other, tossing a coin at each page to decide which original to take the page from and which to throw away." But buying two copies of the *same* book would merely result in scrambling the order of what's in the book. No new information would be introduced. In order to read a book by the *meiotic, sexual* method, you'd have to buy two *separate* books (as if a "male" book and a "female" book), and then randomly combine the pages from both books. *It's that never-previously-existing process of sharing separately-gendered DNA information which natural selection simply can't account for.*

The insoluble problem of overcoming an absolute dependency on novel information where none has ever previously existed is underscored in a delightful discussion of time travel in Paul Davies' fascinating book, *About Time—Einstein's Unfinished Revolution*. Davies cites the work of Oxford physicist, David Deutsch, who, having carefully studied mind-bending time travel puzzles, points us to the following conundrum:

> Consider the example of a time traveler from 1995 who visits the year 2000 and learns of a marvelous new solution of Einstein's equations, published in an edition of that year's journal *Physical Review* by an obscure scientist named Amanda Brainy. The traveler returns to his own year, armed with a copy of the solution, and seeks out the young Amanda, finding her to be a first-year physics student at his local university. He then sets out to teach her relativity, and eventually shows her the new solution, which she duly publishes under her own name in *Physical Review* in the year 2000. The problem about this little story is: Where did the knowledge of the new solution come from? Who made the discovery? Amanda didn't: she was told the solution by the time traveler. But he didn't make it either: he merely copied it down from her paper in the journal. Although the story is entirely self-consistent, it still leaves us feeling bemused and dissatisfied. Important new information about the world can't simply *create itself* in that manner, can it?[140]

Evolutionist "time travelers" who excitedly race ahead of themselves claiming they've found the solution to the origin of sex by feebly offering various theories as to *why* sex would have been advantageous to evolution can't seem to find it within themselves to time-travel back to the point when there was absolutely no genetic or DNA *information* from which sex possibly could have derived, for whatever evolutionary advantage might be imagined. Deutsch is dead right: Important new information about the world can't simply *create itself* in that manner!

Information, information, information. Without genetic information, nothing is IN formation.

Two Simple Analogies May Help

Perhaps one or two analogies might help bring clarity about the unbridgeable difference between mitosis and meiosis. Suppose, for instance, you wanted to copy the very page you're reading at this moment. So, you take your book to a copy machine, lay the book face down on the scanner, push the lid down on top of the book, and press the Start button. The eerie greenish light scans from side to side beneath the book; the copier makes its usual whirring sound; and out the other end of the machine comes your copy. Given the curvature of the book's spine, you might see a slightly distorted image of the page, and maybe even a slight smudge of ink, or perhaps a large black area where no part of the book was in the copying field. But you're happy enough, because you got what you wanted: an exact copy of these very words.[x]

Now suppose you want to make another copy, so you take the copy you've just made (smudges, black areas, and all), and run it through the feeder. Again, there may be additional smudges or other glitches which—if you repeated the same process a hundred times (always copying the latest copy) might look quite different from the original page in this book. Even so, what can you confidently expect to remain constant throughout even a million copies? These very words, right? The copy machine itself, programmed as it is to duplicate exactly what it is given, is incapable of changing the *contents* of that which it duplicates. Right?

How surprised would you be if, as you reached for the millionth copy, you discovered only half of four original paragraphs on the page? And while you're dumbfoundedly examining that page, you notice the copy

---

[x] Like copy machines themselves, analogies like this eventually break down. But it may be worth the risk here to help make the point.

machine is spitting out yet another page which has a single paragraph, the combined size of the two paragraphs on the first page but with a mixture of words that are complete gibberish, as if all the letters had been put into a blender. And suddenly here comes yet a third page which once again has four paragraphs—this time easily readable and ever so interesting (perhaps even with some recognizable similarities to the words you've copied from this book), but certainly not anything like an exact duplication. Oh, and—just for fun—let's say that all three pages are in brilliant color, not black and white!

Admittedly it's an overly-simplistic picture, but the improbability of all of that happening is something in the order of what would be required to go from simple "black-and-white" asexual duplication to "full-color, high-definition" sexual reproduction.

Or we could also think in terms of copying and pasting. For purposes of illustration, I'm going to type this additional sentence. [Done!] Now I'm going to position my cursor at the beginning of the paragraph and "copy" the two sentences. [Done again!] And now I'll "paste" them in after this sentence (highlighting them in italics). *Or we could also think in terms of copying and pasting. For purposes of illustration, I'm going to type this additional sentence.* [Again, done...as you can see.] It's straightforward and easy peasy. No prizes for guessing that this is a simple picture of asexual replication: copy and paste; copy and paste; copy and paste.

The question is: How possibly could a never-varying process of "copy-and-paste" duplication end up being sexual reproduction whereby what is "pasted" is not what you "copied." (Imagine such a thing happening out of the blue one day with your computer!) In fact, male/female, meiotic sexual reproduction is like a combination of the two sentences, all mixed together rather randomly, then somehow reconstructed to once again appear as a new sentence very much different from the two originals. Could some bright "techie" possibly program my computer to do that? Sure. Indeed, that's the danger in all the clever (intelligently programmed) computer models of evolution pitched our way, as if they are irrefutable proof of historical fact. But in a billion years it's not likely to happen randomly in Nature. And certainly not in a single, crucial, go-or-no-go generation.

After all this, are you still thinking that Darwin's secret sex problem is much ado about nothing? Or that it's merely a scientific mystery that evolutionists one day will solve? See if you still think that way after reading the next chapter.

Recap

- One must be careful not to confuse true sexual reproduction with other forms of reproduction involving genetic exchanges, such as horizontal gene transfer.

- It's not helpful when writers use the term "sex" loosely so as to include processes of asexual replication that don't come close to the kind of male/female *meiosis* in true sexual reproduction.

- Given that anomalous reproduction processes such as parthenogenesis have as much of an unknown origin as normative sexual reproduction, it's not surprising that the scientific community does not point to such anomalies to explain the origin of evolutionary sex.

- Depending on the availability of food, organisms such as yeast may reproduce either asexually or sexually, raising the obvious question about the origin of the sexual part of that cycle.

- The original process of *mitosis* could not have been Googled for the DNA information that the previously-non-existent process of *meiosis* would require in order to get going.

Question: Can you see the danger in assuming that, since there's an amazing variety of reproduction in Nature, maybe those forms might have been evolutionary building blocks leading to male/female meiotic sex, when that process is so distinctive that it has its own unique features and operating DNA?

CHAPTER 6

This Gender-bender Is No Accident

The idea of male and female are universal constants.
—"James T. Kirk"

Arguing evolution's inability to provide the unique separate genders necessary for sexually-reproducing organisms requiring both a male and a female, even if both genders are found within the same organism (hermaphrodites) and despite the existence of organisms that mix their genes together without having identifiable genders.

When it comes to sex, it takes two to tango! That is, unless one's definition of "sex" extends to any mingling whatsoever of DNA from two separate sources, in which case even asexual creatures would be having "sex"—and obviously having it without gender. In this book, however, we're talking about real sex. Real *male* and *female* sex. Real male and female *meiotic* sex, which no asexual mitotic replication ever had. *So, the question is: How does evolution explain the origin of gender: male and female?*

Margulis and Sagan propose that "mating types of nearly all other organisms tend to be much less fixed than those of vertebrates. Nor do there have to be only the two genders 'male' and 'female'.... Eventually a host of different genders and their determining mechanisms evolved."[141] Apart possibly from some ciliates and fungi, that's quite a stretch, but doesn't contribute greatly to the search for the origin of the most ubiquitous form of sexual reproduction.

Authors Martin Daly and Margo Wilson join in the chorus, saying, "The existence of sex does not necessarily imply the existence of sexes.... Such a thing does in fact exist in bacteria. Sexes are lacking, but sex is not."[142] "Sex" in *bacteria*? At least Daly and Wilson acknowledge that there's no gender in asexual bacteria. So where did it come from along the path of evolutionary history?

Listening to Richard Michod, you could be excused for thinking that there's a whiff of bait and switch regarding the evolution of gender:

> Why did evolution invent males and females in the first place? As the biblical story of creation tells, the male came first and the female was created so that he would not be alone. What we know about asexual reproduction and virgin birth in other animals suggests the opposite: that females, *or at least the female function* [my emphasis], came first and can exist without the male function. We will find that males have no biological purpose without females, but females may exist without males; which is, after all, what asexual reproduction is all about. But what is the male and female *function*, what is the purpose of the male and female sex?[143]

Are we talking about *females* existing without *males*, or only "the female function" existing in the absence of either distinctively unique females and males which themselves are the products of male-female reproduction? Is there any evidence that "the female function" in any form of asexual replication gave rise to the first true female and first true male? (Somewhere along the evolutionary path, there would have to be a *first* of each.)

At the very least, maleness and femaleness in meiotic sex is radically different from any supposed maleness or femaleness associated with any reproductive process apart from meiosis. In that sense, sex without gender makes *no* sense. Or simply try having *sperm* and *eggs* unassociated with gender! Can we at least start there?

Even dubious circular reasoning at least acknowledges the usual tie between sex and gender. Consider, for example, this highlighting of the obvious: "Males and females most emphatically would not evolve independently. Sex, by definition, depends on both male and female acting together."[144] The last line is certainly true, even if the first line begs the very question being asked. Gotta love the logic: Since typical sex requires both male and female, surely both prototype gender forms *must* have evolved together! But there's not the slightest suggestion as to *how* that possibly could have happened. We just *know* it because we *need* it. Even more so, we just know it because Darwin's "common origin" hypothesis absolutely collapses without it.

In the midst of all the glaring circularity there's at least this kernel of truth: For *meiotic* sexual reproduction to get off the starting line, both male and female prototypes must of necessity have evolved in tandem. But therein lies the problem: gender is no easy matter. It isn't just a one-off unexpected hitchhiker-gene on a random chromosome. By what mechanism could indispensable gender differential possibly have happened...and happened simultaneously? (There's no use of one evolving before the other.) If all you have to work with is genderless, non-meiotic life forms, we're talking magical alchemy or blind faith, not hard science.

Note Graham Bell's candor about the problem of gender:

> Every student knows that homologous chromosomes usually segregate randomly during the division of the nucleus; no professor knows why. Every layman knows that all the familiar animals and plants have two sexes, but never more; few scientists have thought to ask, and none have succeeded in understanding, why there should not often be three or many sexes, as there are in some ciliates and fungi.[145]

Not much mystery here. Whatever you believe about Noah's ark, "two by two"—as in standard-issue *male* and *female*—persists as solid science. But where did these two sexes come from…?

Ridley's Imaginary Gender Tale

Mark Ridley's discussion of gender in *The Cooperative Gene* is probably the boldest offered by any evolutionist, though he certainly doesn't make an auspicious start of it. "Gender is a universal feature of complex life, and intimately related to sex and Mendelian genetics; but its existence may be an accident." Gender, *mere accident*? Ridley explains:

> The reason why we have these two sets of genes per cell is that complex life on Earth happened to originate in an accidental merger event, roughly two thousand million years ago. Two cells merged into one, bringing two gene sets into one cell.
>
> All modern male-female differences (in so far as they are due to evolution) can therefore be traced to an accidental merger event deep in the past. If complex life had not evolved via a merger, it would not have gender. Gender will be the most puzzling feature of complex life on Earth for our extraterrestrial visitors. They will not be able to understand it until they have read our DNA and reconstructed the merger event that is implicit in its codes.[146]

So, had it not been for the accidental Big Merger of the two cells into one, we wouldn't have gender. Any thoughts about where those two merging cells came from…? And when Ridley says "in so far as they are due to evolution," is he seriously suggesting that gender might have come from something other than evolution? If so, from what?

Ridley's road to the first male and first female is long and winding, but oddly enough it begins with...well... "male" and "female"! As in *mother* and *father*, *son* and *daughter*. Using *sexually-reproducing* algae and sea lettuce as examples, Ridley notes that two non-gendered "parents" produce gametes [sex cells!] which fuse to form the offspring cell.[147] So, let's get this straight. We have "parents" that are non-gendered...who together produce sex cells which produce gendered offspring!

If Ridley's explanation is so scientifically sound, why don't other evolutionists simply cite his theory instead of fleeing in panic at the first mention of the origin of gender? Is it because they, too, recognize the logical circularity and dodgy speculation? (Even Ridley seems to be aware that his theory is a bit "out there," saying at one point, "If some of the arguments in this chapter have seemed a little contorted, we can finish by finding a positive merit in these contortions."[148] Contorted, indeed.

In fairness to Ridley and all the other writers, the "unbounded evolution" hypothesis is not proved wrong simply because evolutionists cannot (yet) explain every detail. There's enough mystery around to challenge everyone. But the unbounded, limitless, upwardly-mobile, microbe-to-man hypothesis IS proved wrong if, as Darwin himself said, it can be shown that a process indispensable to its out-working could not in fact have happened.

Myth and Mystery

In doing my research, I came across a book by Mark Jerome Walters (*The Dance of Life—Courtship in the Animal Kingdom*), which I ordered solely on the basis that it promised to address "such profound and intriguing questions as the origin of two-parent sex." Within the first ten pages of "could have been's" and "might have been's," it was clear that Walters had no definitive answer—simply that,

> ...some 500 million years ago, long before the appearance of the first animals, a momentous change began to occur in the history of life. After a long line of virtually imperceptible changes in individuals, visibly different organisms evolved. A new kind of sexuality had emerged: it involved two

> individuals meeting, not to exchange genetic material, but actually to "mate."[149]

I like Walters' distinction between "mating" and merely exchanging genetic material, something like the difference between "exchanging body fluids" and actually having sex.

Then comes the usual confession (with my emphasis): "But the real truth of exactly why sex has become the all-consuming biological activity of so much of animal life remains a mystery. *Most hypotheses, after all, are little more than educated guesses, which in the end may prove to be as fanciful as the abundance of myths that sought to explain the appearance of sex*"—whereupon Walters cites a number of fanciful Germanic, American-Indian, and Afro-American myths (though curiously not the straightforward, matter-of-fact language of the ancient Jewish creation-myth, "male and female created He them").[150]

Asexual Replication While Sex Was Evolving?

There seems always to be those who wishfully put forward the possibility of some asexually-replicating creature blissfully engaging in exact-copy duplication over many generations while somehow a never-before-seen male or female is gradually evolving within that creature. Think again about what "exact copy" means. If *exact copies* are being made, how does any asexual creature even *begin* the process of evolving a more masculine or more feminine side? Or find deep within its inner self a latent sex gene begging for release?

But to play the game just for a moment, suppose that in Generation One an asexual being has a random mutation which is "exactly copied" in Generation Two. How many *random* mutations would it take to make even the simplest, very first male prototype? Now make that same process happen with some other asexual being, by gradually developing (once again by *random* mutations) even the simplest initial female prototype.

(And bear in mind the normative principle regarding mutations—that typically they are not progressive but recessive.) Nor must we forget that these two completely random processes must happen simultaneously. And happen in near proximity. And be both complementary for mating and compatible for an all-new, complex process of reproduction. Need we go on? Are we not completely off the probability charts at this point?

Are Hermaphrodites the Missing Sex Link?

Invariably, some folks will propose that natural selection first evolved dual-gendered hermaphrodites which, of course, reproduce sexually within a single organism (like some snails and slugs, and, of course, flowering plants); or perhaps point to water fleas (*Daphnia*), the greenfly, and the thrips—all of which are capable of replicating asexually (or by binary fission, like bacteria) and also, on occasion, sexually.

Summarizing his discussion in *The Descent of Man*, Darwin rewinds the tape of evolutionary history back through Old and New World monkeys, to amphibians, to fish-like animals, to "the early progenitor of all the Vertebrata [which] must have been an aquatic animal, provided with branchiae, *with the two sexes united in the same individual* [my emphasis]...."[151]

That's both sexes in a single creature, but where did those two sexes come from...?

If Darwin thought that hermaphrodites were the transitional prototype sexual creatures evolving from asexual creatures, he does not take this prime opportunity to argue the case. Nor does he ever elsewhere address either the origin of gender or the origin of sex, without which there could have been no origin of species.

Sexual variety is fascinating in and of itself (and all the more so because without sex we wouldn't have variety!), but let's not take our eyes off the ball. In the simplest possible terms, along the path of evolutionary history we've got to make a mind-boggling leap from *no gender* to the first two *distinctly differentiated, yet compatible genders.* Somehow, some way, comparatively simple "exact-copy" mitosis has to morph into the radically-different, biologically unique process of "mix-and-match" *male/female* meiosis.

That's meiosis with mutual gene reductions of *precisely 50%* (not 2%, 20%, nor 80%). Even if we could come up with a universally-accepted reason as to why sex would have been advantageous to evolutionary progression, how do we possibly explain the never-before-developed *mechanism* that brought about the precisely 50% part of the conundrum... in *both* the male and the female...*in the same first-paired generation*?

Strawberry Fields For Never

"Origin of Sex Pinned Down," shouts the headline of the brief article featured on the website.[152] Certainly got my attention! The excitement over a strawberry research project is palpable, with potential implications far beyond the humble strawberry. "We all came from hermaphrodites, organisms with both male and female reproductive organs," says the writer. "And though the origin traces back more than 100 million years, biologists have scratched their heads over how and why the separate male and female sexes evolved."

Waves of skepticism are already washing over me, but I read on. "Now, research on wild strawberry plants is providing evidence for such a transition and the emergence of sex, at least in plants. And the results, which are detailed in the December [2008] issue of the journal Heredity,[153] likely apply to animals like us.... The researchers suspect the two genes [featured in the study] could be responsible for one of the earliest stages of the transition from asexual to sexual beings."[154]

If we played "Spot the Problem," how well would you do? Actually, there are two problems, one in each paragraph. In the first paragraph (confirmed in detail by the study being hyped), the question is how the male/female "couple" that usually lives happily together and produces offspring in the same hermaphrodite strawberry came to have irreconcilable differences, deciding eventually to split up and go their separate ways. You might say this is about "chromosome evolution," but clearly it's not from *no sex* to *sex*, only—if at all credible—from *conjoined* (hermaphrodite) sexes to *separate* male and female sexes (*dioecy*, is the technical term) as is the usual case.

In the second paragraph, at last we're finally at ground zero, exploring the crucial transition from asexual to sexual beings. But did you catch the problem? The study's hypothesized explanation of the origin of sex from non-sex is drawn from research on strawberries that (what?) *already have the genes for maleness and femaleness, and already reproduce sexually*! For all the hype, we don't end up with delicious strawberry shortcake, but a disappointing strawberry shortcut.

Darwin himself enjoyed a slice of "strawberry shortcut" when thinking about gender. As described by biographer Peter Brent, Darwin speculated that the lowly *cirripede* barnacle—a hermaphrodite with an incredibly fascinating variety of reproductive options—might be the key to the separation of sexes—

> It had already appeared to him probable that species bearing the organs of both sexes were likely in the course of time to throw up individuals in which the characteristics of one sex were beginning to predominate over those of the other. The logic of survival would then lead to the appearance of individuals whose sexual development was opposite and complementary to that of the first group—"and here we have it," he wrote to Hooker, "for the male organs in the hermaphrodite are beginning to fail, and independent males ready formed."[155]

Even if Darwin apparently was thinking here only about the *separation* of sexes, it's a graphic reminder that, without the clearly problematic transition from asexual to sexual, there would be no sexes to separate. Whereas Darwin himself seemed oblivious to the origin of sex problem, others have borrowed his hermaphrodite argument in aid of trying to answer a problem for which it is distinctly unsuited.

In a nutshell, the "hermaphrodite-strawberry argument" is typical of so much of the circular reasoning that is put forward to "explain" the origin of sex. You can't *assume* sex when working backwards to *explain* sex. Perhaps we've overlooked what DNA stands for. You might think it stands for deoxyribonucleic acid, but it really stands for **D**o **N**ot **A**ssume the very conclusion you're trying to reach! If I may say so, evolutionists need to be far more self-chastening about the widespread use of logical (illogical!) circularity. In science, most especially, you can't have your shortcake and eat it too!

To put it all in plain language, natural selection can bring about some amazing changes, but what it absolutely cannot do is select something which doesn't yet exist. And consider the double irony: Not only can natural selection not possibly explain the origin of gendered sex, but without gendered, meiotic sex there couldn't be the very variety which is the lifeblood of natural selection!

What then explains how the first male and female life forms ever came to be? *Is there not more than a little hypocrisy when evolutionists deride creationists for their "mythical Adam and Eve," yet can't begin to explain evolution's own "Adam and Eve"—the first-ever male and female organisms indispensable to sexual reproduction?*

Recap

- If *exact copies* are being made in the process of *mitosis*, no asexual creature could even begin the journey towards evolving a more masculine or more feminine side.

- In order for *meiotic* sexual reproduction to get out of the starting blocks, both the first male and the first female prototypes must of necessity have evolved in tandem—a scenario too good to be true.

- Trying to explain the origin of gender by pointing to dual-gendered hermaphrodites, which have both male and female features capable of reproducing sexually within a single organism (like some snails, slugs, and flowering plants), is classic circular reasoning.

- Darwin wondered whether separate genders might have come from the "separation of sexes" found in hermaphrodites, but that merely begs the question of where those two sexes came from in the first place.

- Because evolution cannot explain the first appearance of its own indispensable "Adam and Eve" prototypes, the Evolution Story is a non-starter.

Question: If you believe that natural selection conceivably might have produced a male organism and a separate female organism, do you think it could have pulled off that feat simultaneously, with just the right kind of male and female organisms so as to mate and reproduce through a never-before-seen process of meiosis?

PART THREE

Evolutionary Sex in the Bigger Picture

CHAPTER 7

Mind the Gap!

Data, data everywhere, but not a thought to think.
—Theodore Roszak

Clarifying what constitutes a "species." Distinguishing between "natural selection" and "sexual selection." Highlighting the crucial difference between the HOW and the WHY of sex. Cautioning that speculative stories and illustrations are the stuff of fiction, not science.

Visitors on the London Underground will be familiar with the recording heard at various Tube stations where there are potentially-dangerous gaps between the train cars and the platform. "Mind the gap! Mind the gap!" says the recorded warning over and over as passengers disembark and board. Likewise, in evolution theory there are any number of gaps of which critically-thinking persons should be aware, most particularly the crucial sex gaps we've pointed to repeatedly. In this chapter, we'll take a closer look at these "lesser," yet vital gaps in our conversations about evolution, beginning with the word *evolution* itself.

As alluded to in the Introduction, the first gap is between two distinctly different definitions of evolution. Are we talking about "bounded evolution" in which species evolve over time within fixed outer boundaries, or the "unbounded evolution" of the romanticized Evolution Story whereby all organisms supposedly evolved from microbe to man?

While the former can easily be observed throughout Nature, the latter is an unobservable, theoretical extrapolation from observable evolution. So, do you believe in evolution? Trick question. One can accept natural selection and even survival of the fittest ("evolution") without accepting descent with modification from microbe to man ("Evolution").

Mind the gap! Mind the gap!

What Qualifies as a "Species"?

Indeed, in the previous paragraph we zoomed past yet another conversational gap when referring to *species*. The word *species* suffers from the same imprecision as the word *evolution*, causing headaches and misunderstandings of all sorts. Just when you think we're talking about various major species, like animals and plants (each a separate major grouping which couldn't possibly interbreed), before you know it, we're talking about various kinds of animals, like dogs, cats, horses, and humans (all different "types" of animals which, again, can't possibly interbreed), then just as quickly about various kinds of dogs (like terriers, poodles, black Labs, and golden retrievers) most of which *can* interbreed.

So, is there evolution within species? It's another trick question. What kind of "species" are we talking about? A *"genus"* (inclusive of all kinds of cat "species" whether kitty cats, lions or leopards)? Or a broader *"family"* or *"phyla,"* inclusive of felines (cats), canines (dogs), and mustelidae (weasels)? And into which of these categories of "species" are we to place the infamous fruit-flies (*Drosophila willistoni*), comprising no fewer than a dozen closely-related forms that could pass for siblings but can't breed together?

The most we can say with any assurance is that there are obvious lines in the sands of mating and sexual reproduction which allow for variation on theme ("new species" if you insist, but more properly simply "varieties" of a single species), but not breeding outside the boundaries of sexually-distinct species. This latter definition of "species" (not simply spin-off varieties) is the sense in which evolutionists themselves use the term when they make a case for common-descent progression from "lower species" to "higher species" (what I have chosen to call "unbounded evolution").

In Darwin's day, it was heresy to deny the fixity of the species. Hadn't God created each and every separate species, all of which reproduced "after their kind"? How then could there possibly be *new* species? How, possibly, could one species give rise to another "higher" species? What fool would dare suggest that all living species had evolved from a single progenitor species? All good questions, but incomplete questions. Is it possible that there might be variations within those "fixed" species? Ongoing, slowly-changing variations? Even perhaps variations which once could interbreed "after their kind" but no longer can?

When Darwin and others began piling up the evidence for variation within species, unsurprisingly it put the cat among the pigeons. It wasn't long before the battle lines were drawn, predictably (as with most disputes) at unhelpful extremes. The religionists drew their line at absolute *fixity*, period! In defiant response, the evolutionists drew their line at absolute *non-fixity*, period!

Once the battle was engaged, the definition of "species" became the first victim of the conflict. In the heat of the battle, the term *species* itself quickly evolved from solely referencing "fixed kinds" to most often referencing "varieties" within kinds, complicating any attempt to find common ground on which both sides could be partially right. Absolute fixity without the possibility of variation defies scientific observation. Life forms ("species"?) evolve! Yet absolute variation with no fixed, species-specific limitations also defies scientific reality. At some point, "species" have boundaries beyond which there can be no evolution. Properly understood, then, a person could be right in saying "species are fixed and unchanging," but equally right in saying "species are mutable and changing." It all depends on how you're using the word *species*.

The conflicting definitions of *species* become most dangerous when framed in the form of a logical fallacy put forward by evolutionists: If any "new species" arises from evolution, then *all* "species" must have arisen from evolution. But that assumption is true only if "species" is defined in such a way as to automatically exclude "fixed" species. If in fact species with fixed outer boundaries are mutable within those outer boundaries, then simply because we might have a "new species" evolving from the original prototype "species" doesn't rule out the possibility that the original prototype "species" itself has fixed limits beyond which it can't change. In which case, there can be *mutability within fixity*—which is to say that the evolutionary process resulting in a "new species" does not mean that *all species throughout Nature* have arisen from evolution.

To support his Grand Theory of microbe-to-man evolution, Darwin was naturally inclined to blur the lines between species as much as possible. It is not surprising, therefore, that in *The Origin*, Darwin walked a less than taut tightrope on the definition of "species":

> Nor shall I here discuss the various definitions which have been given of the term 'species.' No one definition has satisfied all naturalists; yet every naturalist knows vaguely what he means when he speaks of a species. Generally the term includes the unknown element of a distant act of creation. The term 'variety' is almost equally difficult to define; but here community of descent is almost universally implied, though it can rarely be proved.[156]

Distant act of creation? Hmmm. At least that would solve the origin-of-sex problem....

As Darwin moved forward in *The Origin*, he became bolder in asserting the elasticity of species:

> I look at the term 'species' as one arbitrarily given, for the sake of convenience, to a set of individuals closely resembling

> each other, and that it does not essentially differ from the term 'variety,' which is given to less distinct and more fluctuating forms. The term 'variety,' again, in comparison with mere individual differences, is also applied arbitrarily, for convenience' sake.[157]

Just as *conveniently* and *arbitrarily*, Darwin cavalierly dispenses with the overly-confining term "species" so that he can move ahead unhindered with his theory of minor gradations evolving from one "species" to another, from the lowest to the highest.

Mind the gap! Mind the gap!

Is Sexual Selection Helpful?

Perhaps the most definitive non-starter when it comes to evolutionary explanations for the origin of sex is the theory of "sexual selection," which often is confused with natural selection —undoubtedly because of the common term (*selection*) and also because both forms of selection have something to do with evolution. Yet, the idea of *sexual selection* is distinctly different from *natural selection*. Whereas natural selection has to do with evolutionary survival, sexual selection has only to do with the enhanced ability to compete for attention.

Darwin was particularly fascinated with (and equally flummoxed by) the infamous peacock's tail, which he believed had evolved, not to produce body design as such, but solely to help peacocks have a better profile on Match.com. The end goal of sexual selection is becoming the most desirable male among all the other guys out there, or the most attractive female in the neighborhood. With *sexual selection*, the competition is between others of the same sex, whereas in *natural selection*, the competition depends on the success of both sexes combined.

For Darwin, an animal's sexual organs obviously constituted the primary sexual characteristics. The less-obvious features he called secondary sexual characteristics. Those include the peacock's bright, Technicolor "look at me" tail (good for females selecting males) and deer antlers used both offensively and defensively in the competition for sexual partners (perfect for pugilistic males vying for females).

Peacock tails and colorful plumage in finches may be fascinating, but they are a mere sideshow to the main event. For those who happily swallow the whole enchilada of microbe-to-man evolution, all the talk about sexual selection is the pickpocket's "stall," causing us to turn our heads while we're having our logic lifted. When you're on the 22 tram in Prague, keep your hand firmly grasped on the most valuable question of all: How did sex ever get started? No sex, no variation. No variation, no sexual selection.

Mind the gap! Mind the gap!

The Gap Between "WHY" and "HOW"

In their book *Origins of Sex*, Lynn Margulis and Dorion Sagan alert us to one of the subtler gaps in the "origins of sex" conversation. There is more to say regarding their own particular analysis, but it is refreshing to hear Margulis and Sagan openly challenge the usual approach of the scientific community. They rightly insist that evolutionists should more carefully differentiate between the *maintenance* of sex and the *origin* of sex: "This problem of the *maintenance* of sex (that which keeps animals and plants from becoming asexual) must be clearly distinguished from the problem of the *origins* of sex (the ways in which sex first evolved)."[158]

By the usual line of thinking, the *maintenance* of sex is tied to the *advantages* of sex, which somehow is supposed to explain the *origin* of sex. Naturally, it doesn't. Citing the *benefits* of sexual reproduction tells us virtually nothing about how sexual reproduction came to be.

Ask the question: WHY does sex exist, and you'll get no end of pontificated answers. Theories abound as to why, despite the enormous "costs" of inefficient sexual reproduction compared with patently more efficient asexual duplication, sex would be advantageous to the evolution process. Among the many theories advanced as to why sexual reproduction would be sufficiently advantageous so as to explain its origin are at least six classic hypotheses:

> 1. *The Vicar of Bray Hypothesis,* spawned by Weismann, Fisher, and Muller, but facetiously named by Graham Bell with reference to "an English cleric noted for an ability to change his religion whenever a new monarch ascended the throne...[which] teaches that there may be great advantages of easily and gracefully adapting to changed circumstances."[159] The idea is that sex is selected by groups in a uniform but changing environment because it facilitates a more rapid fixation of favorable mutations.
>
> 2. *The Ratchet Hypothesis,* offered by Herbert Muller, also focuses on selection by groups, but posits that, rather than leading to the fixation of favorable mutations, sexual reproduction makes possible the continual elimination of harmful mutations.
>
> 3. *The Tangled Bank Hypothesis* gets its name from the concluding paragraph of the *Origin of Species,* in which Darwin describes a wide variety of life forms, all vying for limited resources on "a tangled bank." In the survival of the fittest, so goes this hypothesis, the diversity of sexual reproduction has a clear edge over non-sexual replication.
>
> 4. *The Lottery Principle,* advanced by George Williams, analogizes to playing the lottery, whereby you have a better chance of winning if you buy fewer tickets and use different numbers than buying a larger number of tickets and use the same number. (Have you been playing it wrong all along!) Similarly, Williams hypothesized that the variety

resulting from sexual reproduction increases the odds of producing offspring that will survive, compared with asexual duplication whose sameness of offspring might be more vulnerable to changes in environment.[160]

5. *The DNA Repair Hypothesis*, posits that sex began out of the need to repair damaged material in the DNA appearing in the continuous and ongoing germ line that survives generations of births and deaths. Whenever there is damage causing deleterious effects, asexual duplication (strictly copying, as it does, warts and all) lacks the ability to correct mutations, so sexual reproduction with its potential for hiding defects for successive generations came on the scene to rescue the day. (This hypothesis has a great deal of weight to it. Sexual reproduction greatly aids in the maintenance of genomic integrity.)

6. *The Red Queen Hypothesis*, suggested by Leigh Van Valen, and championed by others including Bill Hamilton, seems to be the most popular explanation these days.[161] Recognize where the theory got its name? You're right, from the Red Queen in Lewis Carroll's *Through the Looking Glass* who took Alice on a long run that went nowhere, explaining to Alice: "Now, here, you see, it takes all the running you can do to keep in the same place." (Call it the treadmill theory if you wish.) The idea is that even the fittest life forms must constantly improve in order to survive (particularly when potential competitors are also improving), and it's sex that keeps you fit. (No wonder it's the most popular theory!)

I'm hoping that by now you can spot the problem with those six hypotheses and any other hypotheses one conceivably could come up with to explain WHY sex might have gotten started in the first place. Wholly apart from the fact that evolutionary scientists have yet to agree on the exact WHY of sex, the far greater, insurmountable difficulty is the matter of HOW sex possibly could have evolved from non-sex. It's not theoretical *rationales* for sex that matter, but the *mechanics* of how—by any of those rationales—the transition from asexual to sexual possibly could have happened...on the ground...in real time.

Even if everyone agreed, for example, that the primary purpose of sex was repairing damaged genes, it wouldn't begin to tell us how that particular (incredibly complicated) function ever came into existence. If meiotic sex is able to do a better job of repairing damaged DNA than mitotic replication, what information in asexual DNA possibly could have given rise to a novel, sexual DNA capable of performing a function that asexual DNA itself never could perform?

A final caveat: Beware even when evolutionist writers appear to be making a distinction between HOW and WHY, as does Richard Michod in his book, *Eros and Evolution*. Acknowledging that, "In this book, we are primarily interested in 'why' questions," Michod tosses a bone to the importance of HOW. Yet, listen carefully to the kind of HOW he's talking about:

> In biology, it is useful to separate the two kinds of explanation, the how and the why, while keeping in mind their interdependency.... Understanding why a trait exists often requires understanding how organisms work. The "how" question suggests answers to the why question.[162]

Looking at "how organisms work" in order to understand *why* they work that way is a far cry from asking how those organisms originally came to function in that particular way. *Description is not explanation!* One could carefully study how an automobile operates, and from that study reasonably conclude why cars exist (to be a form of transportation) without ever knowing how the car itself came into existence. Michod's HOW question is not the crucial HOW question of this book, and shouldn't be the sole HOW question being asked by evolutionists when faced with the inexplicable conundrum of sex.

If moving from asexual to sexual never could have happened *in fact*, then there can't possibly be an explanation as to *why* it happened.

Without a scientific mechanism to pull it off, it couldn't have happened even if some asexual life form had every reason in the world to try. WHY is not the issue, only HOW?

Mind the gap! Mind the gap!

Science *Fiction*

As a story-teller, Richard Dawkins has few equals. Dawkins' ability to spin evolutionary yarns epitomizes an elite cadre of his evolutionist colleagues (including Charles Darwin himself) who seem to have a special gift of combining encyclopedic knowledge of Nature with a vivid imagination and brilliant writing skills. Yet, it's the *imagination* part that warrants caution, as in the following statement. "The account of the origin of life that I shall give is necessarily speculative," Dawkins admits, but "the simplified account I shall give is probably not too far from the truth."[163]

To give but one of many possible examples of creative envisioning offered by Dawkins and other gifted science writers, consider an excerpt from Dawkins' best-seller, *The Selfish Gene*. Here Dawkins speaks of the struggle for existence among what he terms original "replicators" in the far-distant past of evolutionary history [with my emphases]:

> They did not know they were struggling, or worry about it; the struggle was conducted without any hard feelings, indeed without any feelings of any kind. But they were struggling, in the sense that any mis-copying that resulted in a new higher level of stability, or a new way of reducing the stability of rivals, was automatically preserved and multiplied. The process of improvement was cumulative. Ways of increasing stability and of decreasing rivals' stability became more elaborate and more efficient.

> Some of them *may even have 'discovered'* how to break up molecules of rival varieties chemically, and to use the building blocks so released for making their own copies. Proto-carnivores simultaneously obtained food and removed competing rivals. Other replicators *perhaps discovered* how to protect themselves, either chemically, or by building a physical wall of protein around themselves. *This may have been* how the first living cells appeared.[164]

Even given literary license...*May have discovered? Perhaps discovered*? And, just imagine, all this amazing detail about revolutionary microbiotic discoveries by unintelligent replicators happening *four thousand million years ago*! With enough suppositions along the way, anyone can "prove" anything. "If pigs had wings..." and so forth.

Why is it that unbridled speculation from a fool is sheer fantasy, but unbridled speculation from a scientist is nigh unto fact? (As noted previously, no aspersion is being cast on the legitimate speculation, hypothesizing, modeling, and "What if?" imagining of scientists trying to tease out better understandings of evolutionary processes, only the Dawkins'-style "wild blue" imaginings of popular evolution writers and fantasy-laden documentaries.)

In the context of this book and its focus, what's most interesting about Dawkins is that he boldly hypothesizes in the most clever, creative way (fabricating his story from a string of bald assumptions as long as a strand of DNA) what *probably* happened in the millions of years prior to his beloved "selfish genes" coming into existence, but he doesn't even bother to attempt his patented "Suppose, Suppose, Suppose magic" (where you can simply *suppose* anything you need) to get sexual reproduction from asexual reproduction.

I wonder if Dawkins, the Master Story-teller, intuitively knows he wouldn't get away with his usual fiction even if he tried his hardest to sex-up the "Queen of evolutionary problems." Explaining evolution's stubborn sex problem isn't so easy to do with smoke and mirrors.

Bridging the yawning sex gap is (borrowing Dawkins' own words) *statistically too improbable*

One simply has to smile. Speaking at The Cheltenham Science Festival near our Cotswold cottage, Dawkins was on one of his typical rants as Skeptic in Chief.

> It is pernicious to inculcate into a child a view of the world which includes supernaturalism—even fairytales, the ones we all love, with wizards, or princesses turning into frogs or whatever it was. There's a very interesting reason why a princess could not turn into a frog: It's statistically too improbable.

Undoubtedly, Dawkins (who champions critical thinking) is right about princesses turning into frogs. But are we seriously to think that moving from the copy-and-paste uniqueness of non-sexual reproduction to the distinct uniqueness of sexual reproduction is any less statistically improbable?[y] At least fairytales don't pretend to be factual.

Mind the gap! Mind the gap!

A Whale of a Story!

What do you think? Is the biblical story of Jonah and the "whale" to be believed? If you consider the Jonah story to be mere religious fiction, I can point you to a whale of a story that reasonable people would find no less fictitious, yet is believed by evolutionists with as much fervor as that of the staunchest believer in Jonah's "whale."

---

[y] Ironically, the Grand Theory suggests that over eons of time frogs evolved into princesses. The problem with frogs is that as amphibians they need much more genomic information than humans, since they have to support two radically different life styles. They are, after all, *amphibians*. Of all people, Dawkins, should appreciate that it would be easier for a princess to turn into a frog than the other way around!

I speak specifically here of Carl Zimmer's account of the origin of whales in his book, *Evolution—The Triumph of an Idea*. It's a long, complex story, the simple version being that whales—sharing traits of both mammals and fish—descended from something like primitive cows and hippos, or at least an extinct group of hoofed mammals called mesonychids. Instead of the usual (colorfully-illustrated) story of fish emerging onto land to become mammals, it was just the opposite in the case of whales, as seen particularly in the similarity of anatomy:

> Early whales evolved into remarkably fishlike forms through a gradual series of steps. But inside every whale's finlike flipper there still remains a hand, complete with fingers and a wrist. And while a tuna swims by moving its tail from side to side, whales swim by moving their tails up and down. That's because whales descend from mammals that galloped on land. Early whales adapted that galloping into an otter-like swimming style, arching their back in order to push back their feet. Eventually new whales emerged in which evolution had adapted that back-arching movement to raise and lower a tail.[165]

Galloping? Isn't that something of a fanciful stretch? How by any possible means does Zimmer know they were *galloping*? But more to the point, as always let's get back to the juicy sex bit. That's the real story without which all those intriguing flippers and swimming styles never would have had a chance to evolve. By Zimmer's hypothesis, whales didn't just move from being land-based pre-whales to whales in the ocean. By the theory of common descent, whales necessarily descended from other mammal species, in this case supposedly cow-like or hippo-like creatures (though there are divergent views as to what might have been the actual precursors and line of descent).

Taking just the quickest visuals of a cow (or whatever else it may have been) and a whale, that's a huge gap to fill! So, make a rough guess at how many intermediate forms of mammals such a long evolutionary process would have required. Ten, fifty, a hundred different species?

Thousands? By now you know the drill. At least at each major juncture, we're talking about the need for a sudden and simultaneous new *set* of male and female prototypes, fully equipped to mate compatibly and to reproduce offspring for the next generation.

Let's say we start with the original "cow." What new intermediate species is next? And then what's next? And then what's next? To shorten the process, skip all the way over to the first pair of fully-formed whales. How would those first two whales have mated, and how would they have reproduced offspring? Is that how primitive "cows" would have mated and reproduced? No way. So how do we get from point A to point B *sexually*—intermediate species by intermediate species? Don't gallop over a single one.

Tell us in explicit detail how each separate species mates and reproduces offspring. Tell us what changes in chromosomes and DNA would be required for the discrete sex differences in each generation. Tell us what specialized sexual organs would be needed, what mechanisms for overriding the "pin-number" security systems that keep each species from interbreeding with other species, and what mating habits would have to be developed.

Don't even think of throwing in millions of years on this one. Extra time does nothing for a process requiring every novel organ and instinct in place simultaneously at each step along the way. (With each new species, keep thinking half a penis and half a vagina.)

If you can begin to grasp what a whale of a story that would be, try taking on Zimmer's logic-leaping, detail-omitting sequel: "In a few million years mammals became flying bats; they became gigantic relatives of rhinos and elephants; they became powerful, lion-sized predators."[166] *Fact* or *fiction*? *Science* or *sheer fantasy*?

At least the "whale" that swallowed Jonah would have been a one-time wonder never attributed to a natural process. Zimmer's whale story, requiring literally millions of micro and macro transitional wonders in a non-directed natural world, is exceedingly harder to swallow.

Mind the gap! Mind the gap!

Dr. Jack's Seductive Sex Films

The potential for deception using speculative illustrations, pictures, graphics, and diagrams increases exponentially with today's high-definition videos and slick YouTube presentations. Never is this better demonstrated than by a series of short videos by Jack Szostak ("Dr. Jack") of the Department of Genetics, Harvard Medical School, a distinguished recipient of the 2009 Nobel Prize in Physiology or Medicine. "Dr. Jack's" internet videos are feisty and combative, particularly targeting creationists (whom he ridicules as being, not just unscientific, but downright ignorant) and advocates of Intelligent Design (ID), especially their signature "irreducible complexity" argument.

"Dr. Jack's" evolution videos include a series on the origins of life, and multicellular life, as well as a series on the evolution of the flagellum, the genetic code, and—specifically to the point of this book—sexual reproduction. In 8 short minutes, the viewer is exposed to colorful, action-packed diagrams depicting DNA strands dividing and exchanging places, proteins morphing before your very eyes, a picture of lizards mating, a diagrammatic explanation of how meiosis works, and a diagram of the "ploidy" cycle of alternating diploid and haploid phases.

What follows is a sudden left jab at ID's "irreducible complexity" argument. On the heels of that vitriolic sparring, "Dr. Jack" reminds the viewer that in another of his video clips he has already shown how easy it is to evolve an organ as complex as the human eye, with evolution supplying irreducible complexity in tiny, incremental stages along its way to organs of even greater complexity. So, sex isn't that big a deal.

Flashing quickly by on the screen come diagrams of the supposed evolution from prokaryotic cells to eukaryotic cells; then two single cells fusing; followed by a chart showing how sexual (meiotic) reproduction allows faster replication than asexual (mitotic) duplication, thereby speeding up the process of natural selection.

From that point, it's back to cell fusion and the explanation that cells fused in order to survive environmentally lean times, during which natural selection conveniently deemed it beneficial to give cells an extra copy of DNA in aid of mutation repair, and then—when the good times returned—the cells reverted to a happy haploid state. And with that (*voilà!*) you've got yourself the process of meiosis, complete with homologous chromosomes, chiasma, and recombinant chromatids.

Szostak's whirlwind argument is a combination of sheer speculative assumptions about what must have happened in the deep past—offered as hard fact—in conjunction with "proof" from cleverly-illustrated computer models programmed by a level of intelligence that cannot be shown with any assurance to exist within either Nature itself or natural selection. Invariably along the way, there is the dubious assumption that selective advantage always triggers adaptation and mutation in natural processes, when those processes could not possibly have been prescient enough to know what particular mutations might be advantageous. Szostak gives these processes *minds* before he gives them *brains*. And, per usual, if you can *imagine it happening*, surely it *must have happened*!

Almost unnoticeable with so many non-sequiturs flying quickly by is a concluding slide showing an article from Proceedings of the Royal Society which "Dr. Jack" references as proof of the origin of sex. But it's the easily-unnoticed opening line of the article that steals "Dr. Jack's" supercilious show: *"Despite a great deal of attention, the evolutionary origins and roles of sex remain unclear."* If the glib 8-minute explanation provided by "Dr. Jack" is so compelling, why is the origin of sex still unclear? Why aren't all the other evolutionist writers citing Harvard's esteemed Nobel recipient instead of simply throwing up their hands?

Instead of evading the Queen of evolutionary problems by endless discussion of the *advantages* of sex rather than the *origin* of sex, why isn't Jack Szostak universally hailed as the embattled Queen's brilliant rescuer?

"Dr. Jack's" video clips always close with the tag line: *Think about it!* Good idea. Thinking about it for myself, W.C. Fields' classic quote comes to mind: "If you can't dazzle them with brilliance, baffle them with bull."

Mind the gap! Mind the gap!

Story-Telling, or Telling Stories?

Former senior paleontologist at the British Museum of Natural History, Colin Patterson, fired a shot across the bow of those who would use "the tree of life" story to *prove* evolutionary progression from one species to another:

> As it turns out, all one can learn about the history of life is learned from systematics, from the groupings one finds in nature. The rest of it is story-telling of one sort or another. We have access to the tips of the tree; the tree itself is theory, and people who pretend to know about the tree and to describe what went on in it—how the branches came off and the twigs came off—are, I think, telling stories.[167]

Story-telling is innocent enough until it becomes the "story-telling" of a child telling fibs. As Stephen Jay Gould reminds us, "a plausible story is not necessarily true."[168]

Maybe militant evolution advocates figure that by the rules of improbability the more implausible the stories they make up, the greater the odds that an uncritical public will accept enough of those stories to convince the gullible of the implausible.

Mind the gap! Mind the gap! Mind the gap!

Recap

- In any conversation about origins, one must "mind the gap" between "bounded evolution" (in which evolution can produce varieties, but not higher and higher species) and "unbounded evolution" (in which the evolutionary process is said to have no limits.)

- Failing to carefully distinguish between "species" (meaning varieties) and "species" (meaning truly novel "higher" organisms) has allowed for the false assumption that if natural selection could produce *any* "species," it could account for *all* "species."

- Asking WHY sex (raising speculation as to the advantages of sex over asexual replication) is not the crucial question, but rather HOW sex ever came to be, without which process there would be no sex whatsoever, whether advantageous or not.

- The supposed chain of evolution from land animals to ocean whales is a painfully contrived story proved false by the impossible transitions from one sexually-unique intermediary precursor to another.

- Scientific literature is awash with imaginary stories and graphic illustrations that, far from being hard science, are not simply unscientific, but deceptive science fiction.

Question: Do you trust that you think critically enough to see through the clever fiction of popular evolution writers, and even to recognize when reasoned arguments about WHY sex might have evolved fall short of explaining HOW sex possibly could have evolved?

CHAPTER 8

When the Dog Doesn't Bark

"As long as it remains invisible, it is guaranteed to remain insoluble."
—Margaret Heffernan

Noting the alarm bells that sound when evolutionists (from Darwin himself to authors of current biology textbooks) sidestep any mention of problems related to the origin of sex, or speak only in muffled tones when the subject arises.

In Arthur Conan Doyle's novel, *Silver Blaze*, there is a "curious incident" in which a dog made no noise. As Sherlock Holmes explains to Dr. Watson: "I had grasped the significance of the silence of the dog, for one true inference invariably suggests others.... Obviously, the midnight visitor was someone whom the dog knew well." When it comes to evolution's problem with sex, Sherlock would be quick to tell Watson that if the dog doesn't bark when you'd most expect it, the silence is telling you something important.

You may be curious about the title I've chosen for this book: *Darwin's Secret Sex Problem*. For many scientists, and particularly evolutionists, the problem of evolutionary sex is well known, if all too quietly acknowledged. As we've noted, evolutionists have even christened it the "Queen" of evolutionary problems and launched any number of studies hoping to come up with a viable explanation. It's not surprising, then, that you can find numerous specialist books and scientific articles in technical journals dedicated in whole or in part to discussing the problematic origin of sex. Yet, you'd be amazed how many people—including evolutionists—have never heard that evolution has a serious "sex problem." (And you should see their eyes light up when you tell them about it!)

If left to textbooks and popular science books, the problem with evolutionary sex might as well be marked *Top Secret*. Typically, there's not the least mention of it. Not even a "We don't know, but it doesn't matter." Or, "We've been studying the problem for decades, but we still don't have a clue." Or even simply, "The origin of sex is a deep mystery."

Nothing about this "secret sex problem" suggests any nefarious conspiracy among the scientific community to hide an embarrassing evidentiary lapse. But to say the least, it's a "curious incident," as Sherlock might put it. Why is it that, on the whole, evolutionists have largely ignored the problem? Why isn't more serious attention being paid to something as critical as the millions of missing sex links from one sexually-unique species to another which threatens the bedrock premise of common descent? For those evolutionists who are in on the secret, why haven't they sounded a general alarm throughout the scientific community?

Could it be that this particular secret would be far too dangerous in the hands of a critically-thinking public? Would revealing this secret in tomorrow's headlines and "breaking news" forever put a stain on "Big-E's" upstanding reputation? Indeed, would a full-blown discussion of evolution's sex problem threaten to undo an otherwise elegant theory? One can't help but wonder if that isn't why Darwin never whispered a word about it. It was Darwin himself who first kept the embarrassing "sex problem" a secret.

Darwin's 790-page magnum opus, *The Descent of Man and Selection in Relation to Sex* (1871), may be the supreme example of the dog that didn't bark. Strange, isn't it? Darwin's huge tome is all about sex: Sexual variation. Sexual adaptation. Sexual attraction. Sexual selection.

But there's not a word about...*sex*! That is, *when* it originated, or *how* it originated. If ever there was a book that ought to begin at the very dawn of sexual reproduction, surely it is Darwin's *Descent*. But not a word. Nary a whimper when one would fully expect this dog to bark.

It can hardly be doubted that Darwin knew of a certainty that sex of all types requires both a male and a female. The timing of his marriage to Emma was interesting in this regard, since Darwin was contemplating marriage about the same time he was formulating his views regarding sexual instinct.[169] Male...female? Like horse and carriage, love and marriage....

Under the mentorship of zoologist Robert Grant during his college days in Edinburgh, Darwin was the first person ever to see the male and female sex cells of seaweed dance together.[170] Did Darwin never think back on that experience and ask himself how *male* seaweed and *female* seaweed ever came to be?

And what about those eight years Darwin spent intensively studying his prized Chilean barnacles? We know that Darwin wondered if these barnacles might have descended from hermaphrodites, which themselves evolved (Darwin surmised) through a series of transitional forms until they produced males and females. [171] Elucidating on that, Darwin biographer Peter Brent shares this insight from Darwin's fourth *Transmutation Notebook* (written in 1837-1838) in a lead-up to *The Origin of Species* (1859):

> those organs which perform nearly the same function in both sexes, are never double, only modified, those which perform very different, are both present in every shade of perfection. —How comes its nipples though abortive, are so plain in man, yet no trace of abortive womb, or ovarium, —or testicles in female. —the presence of both testes & ovaries in Hermaphrodite—but not of penis & clitoris, show to my mind that both are present in every animal, but unequally developed.[172]

Did it never occur to Darwin that both the male and the female sex cells of any species—including hermaphrodites—must have come into existence simultaneously in order produce the next generation?

When he first hypothesized his Grand Theory of common ancestry from simple to complex organisms, did Darwin never once question how an asexual being replicating by mitosis possibly could propagate a never-before-seen life form reproducing by male/female meiosis? *How could he not?*

At one particular point, Darwin surely must have ventured close to considering the Queen of evolutionary problems—when questioning what he believed to be the "law of organization" whereby all varieties must be presumed to function in much the same way. As reflected in the fourth of his *Transmutation Notebooks* (relayed by Peter Brent), Darwin...

> nagged at the problems posed by asexual reproduction, since he now based his hypothesis on the changes that might be produced in a mingling of parental characteristics. "My theory only requires that organic beings propagated by gemmation [a form of asexual reproduction found in mosses and similar plants] do not now undergo metamorphosis, but to arrive at their present structure they must have been propagated by sexual commerce." How therefore could asexual reproduction have the same result? It was a fact, he wrote, that "throws a very great difficulty on my theory;" his solution lay in the simplicity of those forms which had continued with asexual reproduction....[173]

Of a certainty, as Brent relays it, Darwin was aware of the problem of gender, and considered it precisely that—*a problem*...

> The development of the sexes from the animal world's asexual progenitor posed problems for anyone attempting to establish the truth of evolution. Underlying similarities, offering clues to how that development might have taken

> place, were therefore of great importance. "In my theory I must allude to separation of sexes as very great difficulty, then give speculation to show that it is not overwhelming."[174]

Did Darwin ever thereafter attempt to explain (even by way of speculation) why—despite being a "very great difficulty"—the "separation of sexes" was not overwhelming? As far as I can tell, Darwin never once directly addressed the problematic transition from asexual to sexual.

As close as Darwin gets to explaining the "separation of sexes" may be this quote in *The Origin* (with my emphasis): "I look at all the species of the same genus as having as certainly descended from a common progenitor, *as have the two sexes of any one species*."[175] Really? Natural selection provided on-time delivery of the first dimorphic yet compatible pair in each of millions of species? But first things first. How did it happen that the common progenitor itself was first gendered? Darwin doesn't say.

What Darwin Ignored

Actually, the *ultimate* difficulty, both for Darwin and all other evolutionists, has to do with how to explain life from non-life. For evolution of any kind to take place, there must first be something to evolve...and it must have come from somewhere. But from where? Peter Brent speaks bluntly in his Darwin biography, saying, "The problem of the Creation...he eventually solved by largely ignoring it. For his purposes, it was sufficient that life existed...."[176]

It seems that Darwin also largely ignored the second most troublesome threat to his hypothesis (the problem of sex), causing one to doubt Brent's laudatory comment about Darwin's supposed objectivity: "Opposition had to be faced and defeated," says Brent, "difficulties had to be overcome, possible objections met. He did not ignore the awkward fact or the contradictory opinion."[177] Maybe not the *awkward facts*, but certainly the *most devastating* fact—like the Doberman Pinscher that barks at everyone but the burglar!

If ever the dog should have barked for Darwin, clearly it must have been when he spoke in *The Origin* about asexual life forms in contrast to sexually-reproducing species which intercross:

> With respect to organic beings extremely low in the scale, which do not propagate sexually, nor conjugate, and which cannot possibly intercross, uniformity of character can be retained by them under the same conditions of life, only through the principle of inheritance, and through natural selection which will destroy any individuals departing from the proper type.[178]

Since Darwin was absolutely right about the crucial inheritance aspect of asexual beings and the probability that slight modifications of an asexual being would usually be destroyed by natural selection, how could he have overlooked the problematic transition from asexual replication to sexual reproduction so vital to his Grand Theory?

Darwin came incredibly close to being confronted by the Queen of evolutionary problems when he discussed the obvious difficulty of early life forms moving from unicellular to multicellular beings:

> Looking to the first dawn of life, when all organic beings, as we may believe, presented the simplest structure, how, it has been asked, could the first steps in the advancement or differentiation of parts have arisen?
>
> ...as I remarked toward the close of the Introduction, no one ought to feel surprise at much remaining as yet unexplained on the origin of species, if we make due allowance for our profound ignorance on the mutual relations of the inhabitants of the world at the present time, and still more so during past ages.[179]

If Darwin's scientific curiosity ever ventured to the next obvious step—the origin of sexually-reproducing beings from asexual beings—he left his readers completely in the dark. Pity that Darwin didn't at least highlight the problem, even if he had no ready explanation. Had he done so, some wag could have written a clever little ditty, like

Darwin was perplexed
By the origin of sex

If Darwin was privately perplexed, he never let on. How could an astute mind as sharply focused as Darwin's on every possible detail of evolution *not* have grappled with the crucial sex link? Since the problem of sex (especially the origin of male and female) is not simply a "great difficulty" but pivotal to his Grand Theory, why didn't Darwin address that difficulty head-on in either *The Origin* or *The Descent*? A blind spot is one thing; ignoring the obvious is quite another.

When Biology Books Don't Bark

Darwin's books aren't the only ones in which the dog doesn't bark when you'd most expect it. Consider, for example, popular-audience books like *Biology For Dummies*, and McGraw-Hill's SAT prep book, *5 Steps to A 5—AP Biology*. In each book, there's the chapter on cells and cell reproduction (mostly the mitosis of asexual replication with a nod to how meiosis in sexual reproduction is different); then several chapters later comes the explanation of sexual reproduction in both plants and animals, culminating with human reproduction. But between those chapters are invisible, almost cell-like walls, with not the slightest acknowledgment of the crucial gap in evolution theory between asexual replication and sexual reproduction. (Of course, there's also the chapter explaining the sweep of human evolution from simple cells to our own highly evolved species, invariably omitting any reference to the embarrassing sex problem that prudishly can't be mentioned.)

As one might expect, within their respective chapters each form of replication or reproduction is carefully and clearly described in minute detail, but (mark this well!) *description alone is not the same as explanation.* And even the explanation of *function* is not an explanation of *the cause* of that function—i.e., how that function came to be.

Oddly, but not surprisingly, in none of the popular, general-audience books yet surveyed is there is any attempt whatsoever to explain how meiosis came on the scene when mitosis was the only game in town. Does the daunting improbability of that ever happening perhaps explain why such books don't want to go down that road? What scintilla of credibility would remain for any author attempting the hocus-pocus magic act such a dramatic transition would require?

Silence About Missing Sex Links in the Fossil Record

And then there's the fossil record and evolutionary sex. The scientific community's typical response to the resounding silence of the fossil record relative to sex is...yet more resounding silence! Consider, for example, this sweeping assertion from *Biology For Dummies*: "Based on the fossil record, paleontologists have established a solid timeline of the appearance of different types of living things, beginning with the appearance of prokaryotic cells (see Chapter 4) and continuing through modern humans."

A solid timeline? Really? No hint whatsoever that there's a huge gap along that timeline, from asexual to sexual? Indeed, *millions of gaps*, considering each point in evolutionary time where sexually-distinct species supposedly arose?

When you go to Chapter 4, as referenced in the quotation, you learn all about the prokaryotic cells—the ones with no nucleus or organelles (as in bacteria), and then their counterparts the eukaryotic cells—the ones with both a nucleus and organelles (as in plants, animals, algae, and fungi). In Chapter 6, we're told that "Single-celled prokaryotes, such as bacteria, reproduce asexually by binary fission...."[180] and then suddenly (here's where the dog should bark!) we're told rather matter-of-factly about the *differences* between asexual reproduction and sexual reproduction, but there's not the slightest mention of any evolutionary *bridge* between them, especially in the fossil record.

In fairness, of course, bones are easy enough to unearth, but not penises and vaginas; and certainly not chromosomes and sex genes. Graham Bell highlights the obvious, saying, "Sex does not fossilize well, and its early history will perhaps always be a matter for speculation."[181] While that obvious difficulty might seem the perfect defense for whatever incompleteness there might be in the fossil record, particularly at the earliest stages, there's still no mention of the countless sex links missing between millions of those upwardly progressing species so essential to evolution. Look for that discussion yourself. You simply won't find it. No barks, no growls, no whimpers.

Giant Leaps...In Complete Silence

McGraw-Hill's *AP Biology* book is almost humorous in its own super-eonic leap through time. It's that visually-affirming graph (on their page 147), complete with diagramed arrows, showing "How Life Probably Emerged," all the way from the earth's atmosphere being formed by [hypothesized] volcanoes releasing [hypothesized] gases into the [hypothesized] atmosphere, which gases [hypothetically] condensed to form seas, from which [hypothetically] came simple organic molecules, then cell precursors known as protobionts, then our familiar friends the prokaryotes and eukaryotes, all of which brings us to a final arrow (denoted "Natural Selection") and this wonderfully pithy conclusion: "Life continues to evolve today."[182] Well, that's one way to leap an unleapable chasm!

One can appreciate that it's not the purpose of an SAT prep book to fill in all the blanks of a given subject, but how can evolution theory be covered in such great detail without there being even the slightest passing mention of the Queen of evolutionary problems? Don't young people on their way to getting a college education deserve intellectually-honest, critical thinking (especially touching on a subject that is foremost on their minds...*sex*!)?

As it happens, once they're actually in college, young people will likely fare no better. Want a college textbook, not just on biology generally but specifically on evolution? You couldn't do better than the lucidly-written, beautifully illustrated text by Carl Zimmer and Douglas Emlen, titled *Evolution—Making Sense of Life*. It's all there—mutations, speciation, sexual selection, the fossil record—the whole nine yards. Everything you ever wanted to know about evolution, right before your very eyes in vivid color! Everything except the slightest mention of the Queen of evolutionary problems. Not a single word, not even in Chapter 11, titled "Sex—Causes and Consequences;" nor, more especially, in section 11.1, with the sub-heading "Evolution of Sex."

After the promising opening line ("A discussion of sex must start with the existence of sex itself."), the discussion immediately turns to unusual forms of sexual reproduction and then various theories as to why, despite John Maynard Smith's "two-fold cost of sex," sexual reproduction has advantages over asexual replication. Despite page after page discussing "the evolution of sex," the crucial evolutionary gap between asexual and sexual is never even hinted at, much less acknowledged or explored.

Not surprisingly, in Zimmer's earlier book, *Evolution—The Triumph of an Idea*, the "dog" growls ever so slightly, but never barks. Barking should have been deafening in Chapter Ten, titled "Passion's Logic—The Evolution of Sex." The only thing deafening was the silence. "Life is a dance of partners," Zimmer begins, "but no list of life's dance partners would be complete without Male and Female." That sounds a promising beginning. I start salivating with eager anticipation.

"As vital as sex may be," Zimmer continues, "Why do males always have small, mobile sperm, while females have giant, immobile eggs? *Why are there males and females at all?* [Italics mine.] The answers are to be found in evolution." Great, here it comes! At long last, evolution's explanation of the origin of gender. But suddenly, Zimmer's big-bark moment disintegrates into a cop-out whimper: "Sex, biologists now suspect, is itself an evolutionary adaptation. It gives organisms a competitive edge over ones that reproduce without males and females."

No, no, no, no, no! Even supposing there is a competitive edge, that edge might explain the *advantage* of male-female sexual reproduction over asexual replication, but doesn't begin to answer the question of *how* "male" and "female" ever came to be. Carefully combing the pages of Zimmer's work, I've found not the slightest attempt to address the crucial issue presented in this book. *Triumph* of an idea? Unless evolution can explain the mechanism for the rise of gender from non-gender, there is no triumph of the evolution idea, only the tragedy of a grand idea in shambles.

Book After Book, With Nary a Bark

In his own encyclopedic tome, *Evolution*, Douglas Futuyma provides the reader with 655 pages of evolutionary polemics without the slightest attempt to confront what Futuyma acknowledges in passing as "one of the most difficult puzzles in biology."[183] With that brief reference, problem dismissed! No ideas. No barking dog. Sexual reproduction just IS!

And then there's George C. Williams' book with the tantalizing title *Sex and Evolution*, which surely, one thinks, will address the origin of sex.[184] But no. In all his discussion about why sexual reproduction would be advantageous compared with asexual replication, and the distinctions between males and females, not a single attempt is made to explore either the origin of gender or to explain the inexplicable link between asexual and sexual. If ever there was a book when the dog ought to bark, this is such a book!

In his book, *Sexual Selection and the Origins of Human Mating Systems*, Alan Dixson never once addresses the origin of *sex itself*, only how sex may have evolved once the sexual ball got fully rolling in primates.[185] Given the potential for endless discussion of any large subject, it's altogether acceptable for an author to limit the scope of his particular narrow interest (as I have done in this book). But it's something of a curiosity that in a book purporting to explain "the origins of Human Mating Systems," Dixson doesn't venture all the way back to the origin of all mating systems.

After awhile, one has to ask why such gifted evolutionists as Zimmer, Futuyma, Williams, and Dixson utter not a single word on the crucial origin of sex when it's absolutely begging for discussion. And what's the appropriate term for such a glaring omission? Mere oversight? Lack of candor? Unscientific? Perhaps something stronger? If natural selection so obviously provided the all-important bridge between asexual replication and sexual reproduction, why not tell us about it? Shout it from the highest mountaintop! Tweet it out! YouTube it until it goes viral! Let the whole Facebook world hear about it and "Like" it! If it really happened, it must have been a marvelous process worthy of re-telling again and again.

The Most Ignored Queen Ever

Worse than *not recognizing* the yawning gap between asexual replication and sexual reproduction is actually *recognizing the gap*—even naming it "The Queen of problems"—yet then acting as if the problem didn't exist! There's such a thing as excusable ignorance, when you truly don't know something; but that quickly becomes culpable ignorance when you jolly well do know it, yet do nothing about it. If the problem of sex is important enough to be given such a royal title, one would think it important enough to have the entire Evolution industry immediately drop everything else and try to figure it out. Why aren't there more scientific articles attempting good (non-circular) explanations? Why not more books on the subject? Why so much tippy-toeing around the difficulty, and speaking in hushed tones?

It's amazing to see how each source of information has its own way of hiding the ball. Consider, for example, how today's ultimate, omniscient authority—Wikipedia—deftly sidesteps the problem: "The evolution of sex contains two related, yet distinct, themes: its origin and its maintenance. However, since the hypotheses for the origins of sex are difficult to test experimentally, most current work has been focused on the maintenance of sexual reproduction...." then quickly moves on to discuss various sex maintenance hypotheses—each of which is equally difficult to test experimentally! In a situation where there couldn't possibly be a plausible evolutionary explanation, it seems that any excuse will do to avoid altogether the (sexually-reproducing) elephant in the room.

Mark Ridley joins the consensus in his book, *The Cooperative Gene*, quipping that "Evolutionary biologists are much teased for the obsession with why sex exists. People like to ask, in an amused way, 'isn't it obvious?' Joking apart, it is far from obvious.... Sex is a puzzle that has not yet been solved; no one knows why it exists."[186] Even if evolution has no ready answer, shouldn't Ridley at least have *mentioned* the Queen of problems in his book titled *The Problems of Evolution*?

What greater problem could there possibly be!

After a while, it's as if you're walking up to your local dog pound, and not a single bark is anywhere to be heard. Shouldn't that cause one to wonder what's going on?

What "Those in the Know" Know...And Don't Know

As I said at the beginning of the chapter, the widespread silence surrounding Darwin's secret sex problem is deafening. That systematic silence speaks volumes, and all the more so considering that, behind the scenes, there are plenty of evolutionists who are in on the secret. How they themselves deal with that secret is equally intriguing, beginning with the following passage from Mark Ridley. When Ridley finally comes close to speaking of the Queen of problems (in *The Cooperative Gene*), who suddenly shows up from out of the blue but...God!

> God may or may not play dice in the laws of physics and of chemistry. God did not need to play dice in the simple stages of biology, while life reproduced clonally. But the evolution of complex life required a mechanism of inheritance with an inherently random component. Somewhere between the bacteria and us—perhaps at about the stage of simple worms—God did have to start to play dice. Life started to use a randomizing system of inheritance, and all subsequent complex life forms have necessarily been built using the randomizing, Mendelian procedure to pass genes from parents to offspring.[187]

Stuck with a less-than-convincing explanation from evolutionists? Then thank God for sex!

For my money, the book that makes the most concerted effort to explain HOW sex evolved is Margulis and Sagan's *Origins of Sex*. As I've previously mentioned, their book suffers from an excessively expansive definition of sex, which has a way of soft-pedalling the radically-different nature of male/female meiotic reproduction.

Given their expansive view of "sex," perhaps Margulis and Sagan are justified in their wide-ranging discussion of every form of genetic mingling and transfers, whether in bacteria or amoebas, whether asexual or sexual, whether mitotic or meiotic. Indeed, they've done serious work in exploring the intricacies of microbiotic life in all its myriad forms. Few authors have dug as deeply. It's only when they venture into the realm of hyper-speculation that one begins to feel less secure with what they like to call their "narrative" of evolutionary history.

At least they are candid about their methodology:

> In evolution, as in criminology, one is never absolutely sure about a given reconstruction. Nonetheless, it is our pleasure here to provide a scenario for the origins of sex that we feel is consistent with the mass of circumstantial evidence so far accumulated.[188]

I can appreciate a case built on circumstantial evidence, although my judgment here is that their case falls far short of being proved beyond a reasonable doubt—or, for that matter, even by a preponderance of the evidence. Were we at trial, I'm afraid I would often be heard saying, "Objection, Your Honor, calls for speculation." Particularly is that true the closer the authors come to the crucial gap between mitosis and meiosis. Suspicion is aroused when you begin to hear an unusual number of "if's...," and "may have's," and "one could imagine." Reasonable doubt starts to creep in with statements like,

> ...meiosis may have evolved as a further error-correcting device for differentiation in the microbial community we call a cell. We may never know the best explanation for the origin of reproduction and sexuality. We can only state hypotheses and explore the consequences.[189]

For a book exploring the hard science behind evolutionary sex, the language [with my emphases] engenders no confidence that their explanation is anywhere near being reasonable scientific conjecture:

> *It is easy to imagine* many *possible outcomes* of a cannibalistic encounter between early protists. The devoured protist, resistant to its own enzymes, *might* also be resistant to the enzymes of its predator. It *might* then have survived *for any*

> *number of reasons. No doubt* in some cases there would have been genetic differences in a cannibalized pair. *If* these genetic differences, no matter how small, conferred on the grisly couple an advantage over the uncoupled conspecifics, the doubled form would persist. Such a doubled form *might* outcompete its single neighbor because it would tend to have less surface area per volume and *perhaps might* have been that much more able to tolerate desiccation or starvation. The result *would have been* a fused protist cell, incipiently diploid, the hardy product of indigestion (ingestion followed by subsequent failure of digestion). *If, however*, conditions had been optimum for the haploid, such a duplex cell, *most likely* derived from a healthy haploid, would be better off in its original haploid state. *Should* the harsh conditions that spurred diploidy change, selection pressures *would tend* to return the diploid to its former haploid state.[190]

So, let's get this straight....

> IF a random protist devoured another random protist, complete with its DNA;
> And IF the devoured protist was resistant to its predator's enzymes and survived;
> And IF the combined genes of the two protists were advantageous over being uncombined;
> And IF that combined arrangement were selected by natural selection;
> And IF those protists coexisted over time as a double-formed entity;
> And IF that entity out-competed other single-formed protists so as not to starve;
> And IF that double form persisted in an ongoing state of diploidy under favorable conditions;
> And IF that double form sometimes reverted to haploidy with a change in environment;
> And IF that double form persisted in fluctuating between haploidy and diploidy;
> THEN, we have ourselves a process of meiosis!

The many suppositions and hypotheses offered us are certainly interesting and worthy of a thoughtful ponder about fascinating theoretical possibilities. But it's something like a long bank of individual toggle switches—all of which would have to be flipped correctly to have the desired outcome. A reasonable explanation for any given part of an infinitely complex process gives no assurance of a reasonable explanation of the whole. Flip even one of the crucial toggle switches the wrong way, and the circuit doesn't work.

For anyone who puts stock in "Occam's razor," the plethora of desperately-imagined assumptions required to explain the origin of meiosis by time-chance-mutation militates against giving any credibility to Margulis' and Sagan's mega-hypothesized story of sex. Occam's razor states simply that, among competing hypotheses, the one with the least assumptions should be selected. By that test, Margulis' and Sagan's hypothesis should not be selected, but soundly rejected. Further compounding the trustworthiness problem which Occam's razor is meant to address, hardly any of the multiple assumptions can be tested scientifically. (And to think that the simplest alternative hypothesis for the origin of sex—requiring but a single, logically plausible assumption [purposeful creation]—is dismissed outright by the scientific community because it isn't scientifically testable!)

Indeed, "Hitchens's razor," popularized by Christopher Hitchens, (as well as a variant by Richard Dawkins) is equally applicable to the discussion: "What can be asserted without evidence can be dismissed without evidence." Any conclusion based largely on unproven assumptions is in essence an assertion without evidence, which should be dismissed on that basis alone.

Margulis and Sagan get honorable mention in yet another book that promises more than it delivers. In *The Major Transitions in Evolution*, co-authors John Maynard Smith and Eörs Szathmáry devote an entire chapter to "The Origin of Sex and the Nature of Species." For specialists, there is much talk about diploidization, endomitosis, syngamy, haploidy, one-step and two-step meiosis, "sister-killer alleles," and anisogamy. Heady stuff, and certainly fascinating, but all the sophisticated scientific analysis never quite comes to grips with the crucial problem of how evolution explains the major transition from asexual replication to fully-gendered, counter-intuitively-meiotic sexual reproduction.[191]

Most striking for the non-specialist are the oft-repeated admissions about how problematic the various hypotheses cited actually are. For example, "Although conceivable, the explanation is not attractive." And, "As always, the facts are messy." And again, "Although this is the generally accepted view, it is perhaps not quite as obviously true, either in theory or empirically, as one could wish." And again, "Although ingenious, the idea has its difficulties."

I admire the candor, but could wish that the someone, someday might ask the truly hard question: Is it possible that we'll never solve evolution's sex problem because what we're trying to explain could not possibly have happened by evolution?

The One Who Crowned "The Queen"

Of all the evolutionists who have studied and written about the phenomenon of sex, one of the most well-informed is Graham Bell (James McGill Professor at McGill University in Montreal and author of *The Masterpiece of Nature: The Evolution of Genetics and Sexuality*). Bell readily acknowledges the problem, and even provides the headline: "Queen of evolutionary problems." In the opening words of Chapter 1, Bell begins his book stating candidly:

> Sex is the queen of problems in evolutionary biology. Perhaps no other natural phenomenon has aroused so much interest; certainly none has sowed as much confusion. The insights of Darwin and Mendel, which have illuminated so many mysteries have so far failed to shed more than a dim and wavering light on the central mystery of sexuality....[192]

From that opening line, it is already obvious that Bell is not going to be solving the sex problem. Noting that it was probably with the rise from prokaryotes to eukaryotes that mitosis and meiosis first arose, Bell supports the scientific consensus that asexual replication would have preceded sexual reproduction. "It is a brute fact," says Bell, "that, of the major taxa which are nowadays wholly or largely asexual, almost all are protists; and if this has any value at all as comparative evidence, it must indicate that asexuality is the primitive and sexuality the derived state of eukaryotes."[193]

At best, Bell speculates that "it seems eminently plausible that the first sexual organisms were haploid isogametic protists whose sex was not costly," but in the end Bell admits, "we can only guess at the origin of sex."[194]

For all his scientific acumen and prodigious effort, Professor Bell never comes close to solving the fundamental difficulties (nor, humbly, ever claims such a glorious achievement). And to think that Bell's discussion is one of the best around. To this day, the reign of "the Queen" is as secure as ever, and Darwin's Grand Theory, consequently, is as *insecure* as ever.

Bottom line? If there is no initial, original transition from asexual (mitotic) replication to sexual (meiotic) reproduction, it's the end of the fairytale bedtime story of human evolution, not to mention all the other millions of sexually-reproducing plants and animals.

Whether dogs aren't barking when you'd most expect them to, or whether they *are barking*, but barking up the wrong tree, it's elementary, Watson: Something incredibly important is terribly wrong!

Recap

- For all the attention Darwin gave to natural selection, sexual selection, and sexual reproduction, he never addressed head-on the critical origin of sex itself.

- Whether it be in biology textbooks, or chapters or books on evolution, the origin of sex problem is rarely acknowledged; and the separate problem of how natural selection possibly could have provided the first compatible male/female pair of each sexually-unique species is *never* addressed.

- In the rare instance when the "Queen of evolutionary problems" is discussed, invariably it is sidestepped with the question of why sex would be advantageous rather than the question of how male/female sexual reproduction possibly could have evolved from non-gendered asexual replication.

- Attempts at hazarding a guess regarding the origin of sex invariably end up being a speculative series of "what if's," and "might have been's," and "one can imagine"—yet always leading to the predictable conclusion that "the origin of sex is a mystery."

- The fact that the origin of sex is rarely allowed out of the closet for honest discussion speaks volumes about its potential for undermining the credibility of the vaunted Evolution Story.

Question: Before reading this book, had you heard about evolution's sex problem? If not, why do you suppose this problem is being kept so hush-hush when among the scientific community it's quietly known to be the "Queen" of evolutionary problems?

CHAPTER 9

If We Had Any Eggs...

Chickens are slow in coming from unlaid eggs.
—German Proverb

Cautioning against the practice of evolution writers who blithely assume whatever they need in order to build a house of cards supporting a precarious theory.

When I was a youngster, a funny quip was often used to explain the depth of one's poverty: "If we had any ham, we could have ham and eggs, if we had any eggs." It strikes me that this expression well articulates the problem of the interdependency required when it comes to sexual reproduction...and especially gender. With but the slightest twist, we could say: *"If we had any sperm, we could have sperm and eggs, if we had any eggs!"*

So many IF's, so little time:

> If we had a female, we could have sex, if we had a male.
>
> If we had the perfect organs for mating, we could have sexual reproduction, if we had the perfect organs for reproduction.
>
> If we had a process of meiosis, we could reduce our chromosomes precisely by half, if each partner had the same number of chromosomes to start with.

> If we had sexual reproduction, we could have the variety needed for natural selection, if natural selection could have produced sexual reproduction in the first place.

Joris Paul van Rossum takes us deep into a similar problem at the super-complex microbiotic level:

> In organisms, we find an interdependency between DNA and proteins, as DNA does not replicate and translate (via RNA) into proteins without proteins themselves, and proteins do not come into existence without DNA. As replication is accomplished in modern cells through the cooperative action of proteins and nucleic acids, this poses challenges when one wants to reconstruct the origin of life and determine what the first self-replicating molecule was, a molecule that served as both information and function, both genotype and phenotype. No RNA has been found yet that could catalyze its own replication [Ridley 2004: p. 530].[195]

Speaking less technically, *it's all about the importance of interdependence.*

Just look in your garden where busy bees are buzzing about, facilitating the process of pollination. Bees need flowers; flowers need bees. And what if there were no common ordinary grass? We take grass for granted, but without grass, what would grass-eating animals eat? For that matter, without *other* animals, what would carnivores eat? Without an ocean, river, or pond having a fully-stocked food-chain comprised of all sorts of sexually-reproducing creatures, how would fish survive?

Of necessity, each of these sexually-reproducing life forms must have evolved, if not simultaneously, at least in cohorts. Not an evolving branch here by itself, or an evolving branch there by itself. Together! In community! Even the predator and the prey find themselves locked in mutual interdependence. Are you into ecology? If so, my argument is your argument. Sex of *any* type needs sex of *every* type. Nothing exists in a vacuum.

Darwin was impressed by the ecological networking of Nature, though seemingly oblivious to the implications for the separate, independent sexual development of each intertwined life form:

> The mistletoe is dependent on the apple and a few other trees, but can only in a far-fetched sense be said to struggle with these trees, for, if too many of these parasites grow on the same tree, it languishes and dies. But several seedling mistletoes, growing close together on the same branch, may more truly be said to struggle with other fruit-bearing plants, in tempting the birds to devour and thus disseminate its seeds.[196]

The capacity for sexual reproduction, and more importantly the *mechanics* for doing so, could hardly be more different in apple trees, parasites, plants, and birds. Yet for any of them to thrive, most all of them have to exist. And for any one of them to exist, the mechanics of sex had to be fully in place—all at the same time.

The sex problem should have struck Darwin on the head time and time again, as when he spoke further of Nature's complex interdependence:

> Let it also be borne in mind how infinitely complex and close-fitting are the mutual relations of all organic beings to each other and to their physical condition of life; and consequently what infinitely varied diversities of structure might be of use to each being under changing conditions of life.[197]

The "structural diversity" of which Darwin speaks could not possibly have happened without the origin of diversity itself—which is sex. No sex; no diversity. No diversity; no "mutual relations of all organic beings." *Sex, diversity, and harmony. But the greatest of these is sex.*

Sexual Symbiosis

Zimmer and Emlen remind us of the interdependence of Nature when relaying to us the results of pollination experiments on *Rhabdothamnus solandri*:

> These results suggest that native flowers on the mainland of New Zealand are suffering pollination failure because they have lost their pollinators. Because *R. solandri* is a slow-growing plant, it may take a long time for it to disappear from the mainland. But without its coevolutionary partners, it may be doomed.[198]

What were the crucial pollinators? New Zealand's native bird species, almost half of which have been made extinct by the introduction of cats and other mammal predators.

It would be difficult to come up with a better example of plants and animals working together symbiotically. Whereas the evolutionist can talk in evolutionary terms about the problem when coevolutionary partners no longer exist, the even greater problem is explaining how the partners ever came to exist in the first place. The mating and reproductive processes of flowers and birds couldn't be more disparate. Did the sexual apparatus of flowers and the sexual apparatus of birds appear simultaneously, and simultaneously in Generation One of each species?

If the sexual apparatus and processes of a particular flower are dependent upon pollination from a completely separate species (say, a bird), then without Generation One of the pollinating bird, there won't be Generation Two of the flower. Can "coevolution" guarantee simultaneous appearance? Can "coevolution" explain the *sexual leap* from non-flower to flower or non-bird to bird? (Not *why*, but *how*. Not the *benefit* of some new form of sexually-reproducing species, but the *actual mechanics* of organizing everything necessary in both the male and the female forms of that new species so that everything falls neatly into place in that crucial first generation.)

The reality of what must have happened *in fact* at some definitive, make-or-break point along the supposed evolutionary timeline abruptly shrinks the question asked by Zimmer and Emlen: "What effects does coevolution have on diversity over the course of millions of years?"[199] Millions of years simply aren't useful for the interdependent mechanics of sex to be set in motion.

Even Darwin recognized that, given the interdependence of life forms in Nature, evolution doesn't exactly have the luxury of time it often demands.

> Though Nature grants long periods of time for the work of natural selection, she does not grant an indefinite period; for as all organic beings are striving to seize on each place in the economy of nature, if any one species does not become modified and improved in a corresponding degree with its competitors, it will be exterminated.[200]

Would that Darwin had recognized even more so the crucial "Generation One" problem when it comes to having radically-different organs for mating and sexual reproduction.

A Host of Problems

Moving from sexually-reproducing plants and birds to sexually-reproducing *hosts* and their sexually-reproducing *guests*, we see an even closer mutualism putting increased strain on the necessity for simultaneous sexual development. If you're scratching your head wondering what I'm talking about, just make sure it's not lice you're scratching! Lice depend on hosts for survival. It may seem nit picking, but the sex life of lice never would have worked without a host to which the female louse could attach her eggs using a kind of Super Glue saliva compound.

There is such an intimate relationship between the louse and its host that there could have been no gap in the timing of their sexual development. It would have been daunting enough for a *non-bird* to develop the capacity for sexual mating and reproduction *as a bird* (in both male and female versions simultaneously). It would have been doubly daunting for some pre-louse prototype to suddenly develop the mating and reproductive capabilities of a fully-developed louse (in both male and female versions simultaneously). As if that weren't sufficiently improbable, it would have been exponentially even more improbable for the louse to require the existence of an altogether different species for its own functional well-being!

If we had any birds, we could have bird-inhabiting lice, if we had any lice!

How Darwin's Familiarity Bred Contempt

The most fallacious circular reasoning of all comes when you assume virtually everything within the totality of our world while trying to prove the origin of any part of it. That kind of blinkered thinking led to Darwin's blind spot regarding evolutionary sex. Fortuitously *given* a world in full ecological balance—including, especially, sexual reproduction—Darwin had the luxury of hypothesizing how organisms, individuals, and groups might have evolved within such a beneficent ecology. Given enough balance and stability on the planet, then perhaps life forms existing within that already-nicely-functioning biosphere could gradually evolve within the limits of natural boundaries.

Take, for example, these words from the Introduction to *The Origin*:

> It is preposterous to attribute to mere external conditions, the structure, for instance, of the woodpecker, with its feet, tail, beak, and tongue, so admirably adapted to catch insects

> under the bark of trees. In the case of the mistletoe, which draws its nourishment from certain trees, which has seeds that must be transported by certain birds, and which has flowers with separate sexes absolutely requiring the agency of certain insects to bring pollen from one flower to the other, it is equally preposterous to account for the structure of this parasite, with its relations to several distinct organic beings, by the effect of external conditions, or of habit, or of the volition of the plant itself.[201]

When Darwin went *macro*, he overlooked the *micro*, including the no small matter of sexual reproduction. If you *assume* sexual reproduction, maybe, just maybe, a theory proposing common descent could work. After all, descent implies change, and change comes from variety, which is the forte of sex. The good news for Darwin was that he didn't have to account for anything he had conveniently assumed. Which possibly suggests why Darwin never attempted to explain the origin of sex. Why should he bother? Nature was absolutely saturated with sex!

On that score, Darwin is in good company with Richard Dawkins, who is content to say dismissively: Sex happens! End of matter. Case closed. Now let's move on, shall we? Obviously, Dawkins is right: Sex happens. But it didn't *just happen*. For sex to *happen*, something had to *happen* to make sex happen!

The grand irony of Darwin's *Origin of Species* is that Darwin never seriously addressed the origin of sex. To play on the theme-line of this chapter, Darwin simply assumed we had both ham and eggs and went about cooking up a theory as to how they evolved into a ham omelet. Someone should have reminded him it's a good idea to have all the ingredients on hand before you start.

The Fallacy of Similar Mechanisms

As it happens, some evolutionists make that exact "ingredients on hand" argument in an effort to prove how "easily" sex could have evolved from non-sex. For example, in building the case for sex being one of "the ten great inventions of evolution," Nick Lane begins by acknowledging the problem of the origin of sex:

> If we all descend from a sexual ancestor, which in turn descended from asexual bacteria, then there must have been a bottleneck through which only sexual eukaryotes could squeeze. Presumably the first eukaryotes were asexual like their bacterial forebears...but all of them fell extinct.[202]

Did you catch Lane's circularity: *Sexual* eukaryotes squeezing through the asexual-to-sexual bottleneck...? Begs the obvious question of how the first eukaryotes (all initially *asexual*, says Lane) somehow magically became sexual. Was that *before* they entered the bottleneck, or *inside* the bottleneck, or as they *emerged* from the bottleneck? (And how very convenient that all the other asexual eukaryotes "presumably fell extinct"!)

There's yet more glaring circularity when Lane says glibly (with my emphasis):

> When sex was 'invented' by the first eukaryotes...there must have been a handful of cells reproducing sexually within a larger population (as it must have done, because all eukaryotes descend from *an ancestor that was already sexual*) the act of sex itself must have conferred an advantage on the offspring of sexually reproducing cells.[203]

Advantage or no advantage, does Lane not realize that he is assuming an "already sexual" eukaryote in the quest to explain how sex was first invented? *Already* sexual is a far cry from *not yet* sexual! The insurmountable "bottleneck" problem lies precisely between those two radically different ways of reproducing.

Without fully appreciating the "Generation One" problem, Lane at least realizes that sex could not have evolved over a great expanse of time (again with my emphasis):

> The question is: if clones were doomed, *could sex have evolved fast enough* to save the day? The answer, perhaps surprisingly, is 'yes!' Mechanistically speaking, sex could have evolved quite easily. In essence, there are three aspects to it: cell fusion, segregation of chromosomes, and recombination.[204]

Simplistically reducing sexual reproduction to only three rudimentary concepts in the highly complex process of meiosis, Lane suggests that the transition from mitosis to meiosis could have been a quick and easy walk in the park....

Cell fusion? No problem, says Lane. All you have to do is lose the cell wall in bacteria and fusion would almost certainly take place. *All you have to do...?*

Segregation of chromosomes? Again, no problem. "It's no more than a modification," says Lane, "of the existing method of cell division, mitosis, which also begins by doubling up chromosomes."[205] Lane then cites Thomas Cavalier-Smith in support of the assertion that "only one key change is necessary to convert mitosis into a primitive form of meiosis—a failure to digest all the 'glue" (technically *cohesin* proteins) holding the chromosomes together."[206] So chromosome reduction resulted simply from an existing mechanism of cell division.

You mean, reduction by *precisely half* the number of chromosomes? In *both* contributing organisms...and *simultaneously*? And that would be the *one key change* necessary to move from mitosis to meiosis—especially male/female meiosis? Is there not the small matter of the DNA information required to make such a move; and the need for crossing over and recombination? When it comes to the multi-faceted complexity of male/female meiosis, it's not just a matter of an easy "one out" in the top of the first (which any one of the three identified factors alone *might* provide), but a game-clinching triple-play in the bottom of the ninth!

Recombination, you say? Still no problem! "The evolution of recombination doesn't present a problem, as all the machinery needed is present in bacteria, and was simply inherited. Not only the machinery, but the precise method of recombination is exactly the same in bacteria and eukaryotes."[207] Whereupon Lane explains that bacteria take genes from the environment—via lateral gene transfer—and incorporate them into their own chromosomes by recombination. For Lane, apparently *any* kind of recombination is tantamount to *all* kinds of recombination, never mind the unique crossing over and recombination involved, especially in separately-gendered meiosis.

So. what's the big deal, Lane asks? "To press the recombination machinery [of asexual mitosis] into a more general role, in meiosis, was surely a formality." Meiosis from mitosis, *a mere formality*? Whatever happened to the Queen of Evolutionary problems, to which Lane himself specifically refers? How could a mere formality be such a persistent problem for evolutionists?

Speaking as cavalierly as Cavalier-Smith, Lane announces his grand conclusion: "And so the evolution of sex was probably not difficult. Mechanistically, it was almost bound to happen."[208]

If that's the case, why does Lane begin his chapter on sex stating the obvious?

> Some of the best minds in biology have wrestled with the problem of sex, but only an incautious minority have been inclined to speculate about its deep origin. There are too many uncertainties about what kind of an entity, or setting, sex evolved in, so any speculation must remain just that.[209]

If the origin of sex was as easy and inevitable as Lane thereafter claims, why the need for such "incautious speculation"? If all the machinery for sex (meiosis) was right there in asexual (mitotic) bacteria all along, why are evolutionists still scrambling for an explanation?

With nothing but *speculative* fusion, *speculative* recombination, and *speculative* segregation of chromosomes on the table, do you still think all the ingredients for meiosis were readily available when natural selection felt the need to cook up sex?

It's sort of like trying to make an omelet using nothing but *speculative* ham and *speculative* eggs!

"Ham and Eggs" in the Plant Kingdom

For anyone not familiar with all the mind-bending symbiosis in Nature, merely consider Charles Darwin's own greatest-ever fascination: carnivorous plants. In his lab, Darwin was mesmerized by plants such as the Venus flytrap (*Dionaea muscipula*), which catches insects and arachnids with a trapping structure formed by the plant's leaves, triggered by tiny hairs on their inner surface. The insects are digested by the plant for their nutritional value, almost as if the plants were animals in disguise. Equally fascinating are the sundews, which catch their prey with sticky substances; and the incredible pitcher plants whose slippery tubes cause insects to slide down into perilous pools where their bodies are dissolved by the plant's enzymes.

More incredible yet, some pitcher plants actually have co-host insects (*unique species found only with that particular plant!*) that aid and abet the plant in its process of stealing nutrients from the trapped insects. Now work out how that felicitous combination ever came to exist by natural selection! The very first prototype of that special plant would need its own exclusive reproductive mode, which would have been unlike any other sexually-reproducing plant before it. Next, our unique insect species would need the same novel gender appearance for the very first male/female pair of that species. And to top it off, in order for their conspiratorial existence ever to get to the next generation of either plant or insect, their mind-numbing symbiosis would have to happen simultaneously! Half a pitcher plant, half an insect, anyone?

If we had plants, we could have plants and animals, if we had any animals.

And here's yet another "Which came first?" question throwing interrogation-level lighting on the supposed random evolution of disparate species. Which came first: the stick insects (with their own inimitable, wildly unimaginable sex life!) or the plants those creatures imitate? While it's easy for evolutionists to say, glibly, "The stick insects evolved to mimic plants in order to have greater protection from potential prey," the question is *how*? How did such precise, interdependent convergence of two kingdoms take place from random occurrences? How did each organism requiring never-before-seen male and female prototypes in both the plant kingdom and the animal kingdom happen? And happen contemporaneously!

If we had any plants, we could have plants mimicked by insects, if we had any mimicking insects.

If Only We Had the *Right Kind* of Eggs!

Nothing quite excites the imagination as fossil finds and dinosaurs—particularly if it's a fossil *of* a dinosaur, and especially if it's a fossil of a "soft" part, such as the dinosaur eggs discovered in the Gobi Desert. As with all eggs having a shell, delicate balance is crucial to the reptilian egg. The shell must be strong enough to prevent accidental breakage, yet fragile enough for the chick to chip its way free. It must also have just the right amount of water so that the chick will neither drown in too much of it or dry out for too little of it. Size and nutrient content must correspond precisely with the size of the embryo at birth, and there must be special membranes allowing the embryo to breathe and to deal with waste products.

All of these factors are crucial for *each egg-laying species*, including the dinosaurs. Merely consider the difference in size (and corresponding egg structure) from fish eggs, to bird eggs, to chicken eggs, to ostrich eggs, to dinosaur eggs. And what happens if, in the process of evolution, this delicate balance is not present in Generation One? No prizes for guessing correctly.

If we had just the right eggs, we could have Generation Two, if we had 1) the fertilization of the egg within the female before the shell begins to harden; 2) all the vital changes in the urogenital organs; 3) development of the perfect "chipping tool" at precisely the right time; and 4) development of the instinct to know when and how to chip out of the cradle. Any single one of these changes without development of all the other changes simultaneously, and we're fresh out of luck. Reproduction is a package deal. *A complicated, virtually inconceivable, all-or-nothing package deal.*

The Great Switcheroo

And we mustn't forget the other side of the evolutionary coin: It's not just a matter of what new equipment must be "turned on," but what old equipment must be "turned off." If, for example, we're talking about evolution from an amphibious creature to a reptile, the whole scheme of amphibian metamorphosis must be scrapped and replaced...immediately. In one generation. The mating and reproductive commands appropriate for amphibians wouldn't be appropriate commands for reptilian mating and reproduction. Any egg or embryo caught in evolutionary limbo would surely die before hatching. So much for Generation Two!

The story is the same for each and every egg-producing species. Between the immediately-prior species and any newly-evolving species, not only must all-new equipment be *added* but every bit of old equipment must also suddenly disappear. Not over eons of time or multiple generations, mind you, but in Generation One. The very first egg of any new species would have to move and set up housekeeping in a new home on the same day, as it were, connecting utilities in the new house while disconnecting utilities in the old.[210]

Looking at the big picture, try to imagine all the necessary interdependent bits that would have to be in place simultaneously to move from a common-origin reptile to the first common progenitor of mammals. Think of these bits as toggle switches needing to be flipped in Generation One, from—at the very least—the inconstancy of blood temperature in the reptile to the constancy of temperature in the first mammal. From the uric acid in reptile waste to the urea of mammal excretion. From the two dorsal aorta of reptiles to the one great artery of that first mammal. From the reptile's scales to the first-ever mammal's fur. From the diaphragm-less reptile to the very first mammal's never-before-seen diaphragm.[211] And—coming back full circle to our discussion of sexual reproduction—we would need a revolutionary supply of mammalian milk that reptiles couldn't possibly provide!

If we had mammalian milk, we could have mammalian reproduction, if we had every other structural and functional feature switched from reptiles to mammals.

Over and over, the problem is always the same: evolutionary sex is not a lone problem standing off in a corner all by itself. Just as sex is by nature social, the evolution of sex requires countless willing partners all working together in concert.

It is problematic enough that evolution can't begin to explain evolutionary sex. Its further inability to explain the host of crucial factors on which reproductive sex relies is as futile as attempting to make a ham and cheese omelet when you have neither ham, nor cheese, nor eggs.

Recap

- Evolution advocacy is notorious for simply assuming whatever is needed to make the Evolution Story work, whether it's the existence of sex itself, or felicitous laws of nature, or the magical coevolution of necessary support systems and symbiotic organisms.
- Ecological interdependence would have required from an unpurposing process of natural selection that a plethora of sexually-reproducing grasses, trees, plants, and animals all evolved in lock-step progression, complete with the sexual mechanisms required for each individual organism within that mutually-dependent ecological landscape.
- Sexual symbiosis such as that between host and guest (as in birds and bird-inhabiting lice), predator and prey, and the unusual insects that conspire with carnivorous plants to trap their evening meal, defies any evolutionary explanation.
- With evolutionists, the proverbial chicken and egg proposition is not "Which came first?" but "If we had any males, we could have males and females, if we had any females."
- Because sex is abundant throughout Nature, Darwin had the luxury of starting off with the sexual reproduction without which his prized natural selection couldn't possibly operate in Nature, but—herein lies the fatal flaw—natural selection could not even remotely have produced the process of sexual reproduction itself.

Question: If you accept "Big-E" Evolution, is it possible that you've been far too blasé in assuming a whole host of collateral assumptions which "little-e" evolution never could have provided?

CHAPTER 10

Extinction of an Unfit Paradigm

The inertia of the human mind and its resistance to innovation
are most clearly demonstrated not, as one might expect,
by the ignorant mass...
but by professionals with a vested interest in tradition
and in the monopoly of learning....
—Arthur Koestler

Examining faith-like evolutionary dogma, especially through the eyes of Charles Darwin; and proposing a reconciliation of opposing viewpoints.

What do you think so far about this book's premise that the problem of sex is fatal to microbe-to-man evolution? Convinced? Unconvinced? If Arthur Koestler is right in the chapter quote above, your answer might depend upon whether you consider yourself part of "the ignorant mass" or part of the elite scientific establishment having a vested interest in Darwin's Grand Theory. I would like to think that the line doesn't fall that neatly between the two, but if Koestler is right, this book will be dismissed out of hand by those who most seriously need to consider it.

Joris Paul van Rossum (himself a believer in evolution's premise of common descent) describes how tentative, novel challenges to accepted paradigms can sometimes lead to a paradigm in crisis:

> Normal scientific research is directed to the articulation of those phenomena and theories that the paradigm already supplies.... In the course of time, some puzzles prove resistant to solutions. Facts appear that cannot be aligned with theory, phenomena identified that do not fit the conceptual box that the paradigm provides. In short, anomalies appear. Eventually, it becomes impossible to deny that there are problems which cannot be solved within the framework that the paradigm provides, leading to a crisis.[212]

Would that such a crisis were looming on the horizon! How long will the Queen of evolutionary problems be kept waiting?

Van Rossum points specifically to the problem of sex as one of those rare, pesky anomalies that not only stubbornly refuses to fit the sacrosanct theory but threatens an entire worldview inextricably linked to that theory:

> The paradigm for which sexual reproduction is an anomaly is not merely the theory of natural selection, but its underlying set of ideas and basic assumptions that serve as the framework through which the biological world is understood and interpreted.[213]

Little wonder, then, that there might be much predictable resistance to the thesis of this book. Evolution's sex problem challenges not simply Darwin's Grand Theory but the broader interpretative framework through which evolutionists see their world, scientifically and otherwise. Falsify the lesser, and you falsify the greater.

Bold, indeed, is any claimed falsification of so widely-accepted a premise as Darwin's Grand Theory. Yet it was Darwin himself who laid down the gauntlet—and even provided the acid test—as explained by Van Rossum with particular reference to the problem of sexual reproduction:

> Testing the theory of natural selection means the observation of living beings and their features and processes: if they

are explainable as evolved through natural selection, sexual selection or genetic drift, the theory stands; but if they cannot be explained, the theory falls. Darwin himself also referred to this method of falsification, when he mentioned in the Origin of Species:

> *If it could be demonstrated that any complex organ existed, which could not possibly have been formed by numerous, successive, slight modifications, my theory would absolutely break down.*[214]

> In other words, natural selection would be refuted if living beings had organs which could not have been formed gradually, which is a necessary condition for it to have been formed through an evolutionary process of adaptation through natural or sexual selection. The very existence of such an organ would imply the falsification of that theory.[215]
>
> The fact that sexual reproduction cannot be explained by the theory of natural selection implies that sexual reproduction is a falsification...and is...a phenomenon that is fatal to the theory of natural selection.[216]

From Darwin's own pen comes the collapse of his prized theory of "unbounded evolution." Because evolutionary gradualism fails to account for the "Generation One" problem in the supposed transition from exclusively asexual organisms to the very first male/female meiotically-reproducing organism (as well as the simultaneous origin of the first sexually-compatible male and female of each new species), Darwin's Grand Theory is well and truly falsified.

Play it again, Sam: "*If it could be demonstrated that any complex organ existed, which could not possibly have been formed by numerous, successive, slight modifications, my theory would absolutely break down.*"

There you have it, in Darwin's own words. No, not Darwin's important contribution to "natural selection" generally; only to his Super Theory:

microbe-to-man evolution. "Natural selection can lead to design, but not to all forms of design; to adaptations, but not to all kinds of adaptations."[217] Darwin's "natural selection" was positively astute; he just exceeded its limits by about a million-fold, making it the classic case of "too much of a good thing." In football lingo, Darwin out-kicked his coverage.

Objectivity: More Proclaimed Than Practiced

In the Preface to his book, *Evolution—The Triumph of an Idea*, Carl Zimmer praises the objectivity of scientists (particularly in contrast to proponents of Intelligent Design.) "In an actual scientific controversy," says Zimmer, "scientists go to major conferences and present their results to their peers, who can challenge their data face-to-face."[218]

Of course, Zimmer speaks only of those controversies that evolutionists agree among themselves to disagree about. What kind of objective reception would any scientist receive if the paper being presented were to challenge the very bedrock of the Grand Theory on purely scientific grounds—or, for that matter, using Darwinian logic to nullify his theory using his own test for nullification? (Naturally, any challenge alleging supernatural origins would be dismissed out of hand as being, by definition, non-scientific—even if supernatural origins are theoretically plausible and thus fair game for even scientific scrutiny.) Just how objective can peer review possibly be if all of one's peers share the very perspective being challenged?

In this regard, it's interesting to note Stephen Jay Gould's observation about scientific objectivity:

> I am...an advocate of the position that science is not an objective, truth-directed machine, but a quintessentially human activity, affected by passions, hopes, and cultural biases. Cultural traditions of thought strongly influence scientific theories, often directing lines of speculation,

> especially...when virtually no data exist to constrain either imagination or prejudice.[219]

One would be foolish indeed to challenge the underlying facts upon which the process of natural selection is based, or even to deny that both natural selection and sexual selection are at work in our world. (Had that been the full extent of Darwin's thesis—that species are not immutable—he needn't have worried that "it is like confessing a murder."[220]) But only a brave soul would show up at a scientific conference daring to suggest that natural selection cannot account for variety itself, and especially for the origin of sexual reproduction which most explicitly provides that variety.

It wouldn't matter that such a challenge to the popularized Evolution Story would leave completely intact the processes of natural selection and sexual selection clearly observable within fixed boundaries of living beings. Every "objective scientist" in the room would quickly recognize the true threat of evolution's sex problem: the vulnerability of the shared, fundamental core belief of the assembled scientific congregation. That core belief is what Gould calls "the simple *fact* of evolution—defined as the genealogical connection among all earthly organisms, based on their descent from a common ancestor."[221] For the "common ancestor" thesis to hold scientifically, natural selection (whether Darwin's gradualism, Gould's punctuated equilibrium, or some other theory) absolutely *must* account for the transition from asexual replication to sexual reproduction, and from one sexually-unique species to another (higher) sexually-unique species.

Evolution's sex problem is not a challenge to legitimate science, but—stubbornly ignored and unanswered—is a challenge to scientific objectivity. Lacking such objectivity, scientific conferences have a way of evolving into edifices with stained glass and steeples.

The Rape of Reason

An internecine battle among evolutionists themselves reveals much about the lack of objectivity on the part of evolutionists generally. As relayed by Carl Zimmer, quite a brouhaha was stirred by Randy Thornhill and Craig Palmer with the publication of their book, "A Natural History of Rape," which posits that rape is an evolutionary adaptation—enabling men to increase their reproductive capabilities by gaining access to women they otherwise couldn't attract.[222] You're kidding! Did Thornhill and Palmer really mean to say that *reproduction* is an unconscious evolutionary motivation for the brutality of rape? No wonder their book received a scathing review by University of Chicago's Jerry Coyne and Harvard's Andrew Berry. But it's what Coyne and Berry had to say about scientific objectivity that grabs my attention.

> In exclusively championing their preferred explanation of a phenomenon, even when it is less plausible than alternatives, the authors reveal their true colours. *A Natural History of Rape* is advocacy, not science.... Thornhill and Palmer's evidence comes down to a series of untestable "just-so" stories [which Zimmer notes is a reference "to the title of Rudyard Kipling's 1902 book of children's tales, whimsically explaining how the leopard got its spots, the camel its hump, and the rhinoceros its skin."][223]

Less plausible than alternatives! For someone like myself who rejects the romanticized Evolution Story, those are powerful words. Evolutionists might dismiss the most obvious alternative to microbe-to-man evolution as unscientific and even risible, but when it comes to explaining the from-out-of-nowhere existence of gender, and male/female meiosis, and sexual reproduction, it has to be said that the simple, if "unscientific," alternative is at least as plausible if not more so.

Would that Zimmer's follow-up were true: "To document real adaptations, biologists use every tool they can possibly find, testing for every possible alternative explanation they can think of." I appreciate that

Zimmer is speaking about normative scientific method. But the plain fact is that scientists don't always consider "every possible alternative they can think of." In fact, like Zimmer, they aggressively denigrate the most reasonable alternative, simply and solely because it doesn't fit their model of what is "scientific." But what if that "unscientific" alternative was actually *true*? (Indeed, if factually true, then it *would be scientific*!) Shouldn't *truth* be the ultimate quest?

And then there's that other phrase packing a wallop: *Advocacy, not science!* What else is it but sheer advocacy when scientists exclusively champion a preferred explanation of a phenomenon, even when there is an obvious Achilles' heel undermining their preferred explanation—a "hot potato" so hot that not even the most articulate advocates of evolution dare touch it! *In such a case, have we not moved from science to advocacy?*

Breaking Through Paradigms

The more I reflect on how so many science-trained specialists can totally ignore evolution's fatal flaw, the more I fear that the problem is not the logic or illogic of argumentation but rather the paradigm through which one filters whatever argumentation is made. Two scientists can evaluate the same phenomenon (or argument)—one being convinced, the other unconvinced. Surely, then, one must ask: Are *a priori* beliefs preventing an objective hearing of the argument? We all have personal paradigms. I do; you do; everyone does.

What's a paradigm? It's any lens through which we filter facts. For creationists, the paradigm is the Bible, which bases its central themes from Genesis to Revelation around the core premise that God is the Creator of all things. Since for atheists there is no God, any notion of supernatural, divine creation is a non-starter. Any explanation for what can be observed in Nature must fit an exclusively materialist, naturalist paradigm. For a Marxist, interestingly enough, one's paradigm might affect the particular way one views how evolution actually played out. Michael Pitman has suggested, for example, that Stephen Jay Gould's theory of punctuated equilibrium reflects his political leanings, saying that "The dialectical laws of Marxist philosophy are explicitly punctuational and Gould believes it is no accident that Russian paleontologists support a model similar to his."[224]

Paradigms can be pernicious when confronted by stubborn facts. Darwin himself might well have uttered those very words when he presented what he believed to be evolutionary fact, knowing full well that his naturalistic "facts" would shatter the theistic paradigm of the many centuries before him. By Darwin's theory, God was no longer the Divine Cause of all we observe in Nature. In the battle for the survival of "the fittest explanations," Divine Cause had fallen prey to Natural Cause. God himself was now extinct, over time to be fossilized in the shale of primitive (superstitious) human history.

From Darwin onward, therefore, scientific observation has *assumed* evolution as the Prime Cause of any biological phenomenon, just as certainly and fundamentally as God had previously been assumed to be the First Cause. Prior to Darwin, scientific observations would have been solely attributed to divine origin. Consider, for example, this observation written in 1702 by Austrian ornithologist Baron von Pernau, cited in Mark Walter's *The Dance of Life*:

> The Wood-Lark eagerly follows the attraction call, in contrast to the Skylark that does not care about it; the reason for that difference probably is that God's inexpressible wisdom... did not implant in Skylarks that method of attracting each other.[225]

Today, we would be shocked to see the slightest attribution to divine cause in any scientific paper. Nowadays, the same sentence would read: "the reason for that difference is that evolution did not implant...." Same observation; different assumed cause. Same observation; different paradigm. Same observation; different dogma.

Just how ironic is it that, as the supernatural has been reduced to the natural, the natural has taken on the irrefutable, incontrovertible trappings of the supernatural?

So, is it *scientific evolution* we're talking about, or *philosophical evolutionism*? Which of those can best respond to the "Generation One" problem? Which will be willing to acknowledge the devastating problem of multiple "black-spot intersections"? Which is more likely to aggressively seek an answer to the indispensable origin of genders necessary for sexual reproduction? Which will be more inclined to seriously engage on the objective issues surrounding the Queen of evolutionary problems? To dismiss logical argumentation solely because of one's faith in Darwinism betrays any claim of being scientific and ends up donning the same religious garb Darwinists so eschew. Blind science is hardly distanced from blind faith.

Much to be hoped for is the open-mindedness of Judith Hooper in *Of Moths and Men*: "For the record, I am not a creationist, but to be uncritical about science is to make it into a dogma."[226] When it comes to evolution's sex problem, no one needs to be a creationist to acknowledge there's a point at which it simply has to be said: "Houston, we have a problem."

Sex, Darwin, and the Supernatural

Since gradualism was an indispensable *sine qua non* of his Grand Theory, Darwin was adamant that life had not unfolded by leaps or jumps into permanent periods of stasis (as is reflected in the notorious fixed boundaries of mating and reproduction throughout the species). In that light, this argument from Darwin is intriguing:

> He who believes that some ancient form was transformed suddenly...will be almost compelled to admit that these great and sudden transformations have left no trace of their action

> on the embryo. To admit all this is, as it seems to me, to enter into the realms of miracle, and to leave those of Science.[227]

It's remarkable that Darwin should couch the options in those particular terms. By Darwin's own concession, his Grand Theory is a high-stakes gamble of the highest order. If gradualism fails to explain the "many structures beautifully adapted to all other parts of the same creature and to the surrounding conditions" (surely an apt description of the unique, species-exclusive sexual anatomy and reproductive function in millions of distinct species), then by Darwin's own logic, the only alternative explanation for that phenomenon is the very realm of the miraculous that Darwin ultimately rejected.

Do you find it interesting that Darwin should so starkly juxtapose science and the realm of the miraculous, and then summarily dismiss the realm of the miraculous as if the two were mutually exclusive? By definition, a miracle is something which cannot be accounted for by science or the *natural*, only by the *supernatural*. But is there only one operating system in the universe—the *natural*? Could there not be *both*—both the realm of science (in which Nature's laws hold sway) and the realm of the miraculous (in which Nature's laws are supernaturally set in place or suspended)?

Of course, Darwin did not automatically associate "Nature's laws" with the miraculous. Indeed, in his "Historical Sketch" introductory to *The Origin*, Darwin pointedly contrasted *law* and *the miraculous* when referring to the work of Lamarck: "He first did the eminent service of arousing attention to the probability of all change in the organic, as well as in the inorganic world, being the result of law, and not of miraculous interposition."[228]

But if Nature's laws were not set in place miraculously, the obvious question is: Where did they come from? Random chance is one thing; *law* quite another—especially a "law" assumed to bring random chance from disorder to order; from chaos to perfection.

I'm certain that this discussion is the last thing Carl Zimmer had in mind when he warned that "scientists have to be on their guard when they try to reconstruct how evolutionary transformations took place, because it is easy to impose a simple story on a counterintuitive reality."[229] But what if that is exactly what evolutionists are doing—substituting a simple (incorrect) story for a wholly counterintuitive, but absolutely true (and therefore *factual*) reality?

Among all the "multiverses" that scientists now talk about, might not this universe itself be part of a much larger, more profound "multiverse"? On what basis does science openly accept the possibility of all kinds of "multiverses" while arbitrarily ruling out one (and only one) particular kind of "multiverse"? Are we to believe that all *material* realities are possible, but no *immaterial* reality is? That all *natural* causes are possible, but no *supernatural* cause is? By what law or logic?

Mere mention of the word "law" ought to have set off alarm bells for Darwin. Laws may regulate the material, but are not themselves material. And Darwin was all about *laws*. "Many *laws* regulate variation," said Darwin, "some few of which can be dimly seen..."[230] And again, "the mysterious *laws* of correlation." And again, "The *laws* governing inheritance are for the most part unknown." And yet again, "the *laws* of embryology."

What a serendipitous luxury for Darwin! His Grand Theory was made possible by a host of mostly unknown but felicitous *laws* operating silently and inexorably throughout Nature. Did Darwin never ask himself what lawmaker had set those laws into place? Or in which realm—necessarily separate from Nature itself—that lawmaker would have been formulating all those beneficent laws?

Darwin's critics were quick to point out that he was speaking of natural selection as if it were an active power or deity, to which Darwin responded, "it is difficult to avoid personifying the word Nature; but I mean by Nature, only the aggregate action and product of many natural laws, and by laws the sequence of events as ascertained by us."[231] Leading to the obvious question: Where did those ascertainable, universal forces come from?

Which would be the greater miracle: directed acts by a purposeful deity acting outside of Nature, or non-directed Nature itself somehow coming up with non-directed direction? If one is forced to introduce "personification" into the conversation, what does that imply if not some force external to Nature itself?

Are the Natural and the Supernatural Mutually Exclusive?

Call me a hopeless Luddite, but at this very moment I am typing in Word, using Windows 7 on a MacBook Air. (I know, I know. Let's just say I'm gradually evolving!) So, I've got two radically-different operating systems (Apple and Windows), but they are happily interfaced by my nifty "Parallels" program allowing me to take advantage of either system with merely the swish of my fingers. Different operating systems needn't necessarily be mutually exclusive.

Sometimes systems *are* mutually exclusive, as in the fundamental distinction between a heliocentric worldview and one that is geocentric. (It can't be true that the earth revolves around the sun and equally true that the sun revolves around the earth.) But other systems *can* lie alongside each other quite comfortably, each with its own function and role to play.

Even with his growing rejection of any miraculous creation of the countless life forms found throughout Nature, Darwin himself seems to allude to this duality, saying,

> The 'Philosophy of Creation' has been treated in a masterly manner by the Rev. Baden Powell, in his 'Essays on the Unity of Worlds,' 1855. Nothing can be more striking than the manner in which he shows that the introduction of new

> species is 'a regular, not a casual phenomenon,' or, as Sir John Herschel expresses it, 'a natural in contradistinction to a miraculous process.'[232]

On one hand is the *natural*; on the other hand, the *miraculous*.

This duality surely lies behind the words of Darwin's wife, Emma, in a letter she wrote to him in 1839. What can be proved scientifically, she urged, should not prevent his accepting "other things which cannot be proved in the same way & which if true are likely to be above our comprehension."[233]

Whether Emma was a motivating factor will never be known, but Darwin himself acknowledges that self-same duality in the closing lines of *The Origin*, saying,

> Thus, from the war of nature, from famine and death, the most exalted object which we are capable of conceiving, namely, the production of the higher animals, directly follows. There is a grandeur in this view of life, with its several powers, *having been originally breathed by the Creator into a few forms or into one* [emphasis mine]; and that whilst this planet has gone cycling according to the fixed law of gravity, from so simple a beginning endless forms most beautiful and wonderful have been, and are being, evolved.[234]

Only an arbitrary, *a priori* perspective prevents one system from easily explaining what another system clearly cannot explain.

Clearly Rejecting the Miraculous

Whatever else Darwin might have thought about the realm of the miraculous, or the role of "the Creator" (which he references in subsequent editions of *The Origin* after the first printing[z]), the last thing he wanted people to believe was that "the miraculous" had anything to do with the evolution of species (necessarily including the elegant, species-specific sexual phenomena Darwin clearly would have observed).

Darwin was candid about his accomplishment in *Descent of Man*, proudly proclaiming: "I have at least, as I hope, done good service in aiding to overthrow the dogma of separate creations."[235]

In closing his Introduction to *The Origin*, Darwin says explicitly:

> The view which most naturalists until recently entertained, and which I formerly entertained—namely, that each species has been independently created—is erroneous. I am fully convinced that species are not immutable; but that those belonging to what are called the same genera are lineal descendants of some other and generally extinct species, in the same manner as the acknowledged varieties of any one species are the descendants of that species.[236]

His closing reference to "the Creator" notwithstanding, Darwin seemingly took pains to make sure no one got the idea that he believed in any kind of special creation. Consider, for example, his caveat: "...it may be well to bear in mind that by the word 'creation' the zoologist means 'a process he knows not what.'"[237] And again, "Why, on the theory of Creation, should there be so much variety and so little real novelty? Why should all the parts and organs of many independent

---

[z] In their book *Evolution—Making Sense of Life*, Zimmer and Emlen chose (on p. 28) to omit Darwin's reference to "the Creator" which Darwin inserted following the first edition. Stephen Jay Gould also omits the "Creator" reference when he presents the Darwin quote in the Prologue to *The Panda's Thumb* (on p. 16). Left for readers to decide is what Darwin meant in the first edition by "having been breathed."

Darwin's biographer, Peter Brent, says (426-427) that shortly after the first edition of *The Origin*, Darwin "set about making minor changes. For example, he attempted to placate the religious feelings that he had outraged by here and there adding the phrase, 'by the Creator', so suggesting the belief that matter had first achieved life through the intervention of some divine agency."

beings, each supposed to have been created independently for its proper place in nature, be so commonly linked together by graduated steps?"[238]

Playing Peekaboo with the Creator

That said, at surprising points in *The Origin,* Darwin reverts to creation language, as when he is discussing how natural selection could produce complex organs such as the eye. In one particularly mystifying passage, Darwin seems in one breath to both affirm and deny the Creator:

> It is scarcely possible to avoid comparing the eye with a telescope. We know that this instrument has been perfected by the long-continued efforts of the highest human intellects; and we naturally infer that the eye has been formed by a somewhat analogous process. But may not this inference be presumptuous? Have we any right to assume that the Creator works by intellectual powers like those of man?[239]

Whatever else Darwin had in mind about the Creator's intellectual powers, at the very least he seems to acknowledge the existence of a Creator, particularly in contradistinction to man. Yet, could it be that Darwin saw natural selection itself as "the Creator"? In his follow-up comments, Darwin virtually endows natural selection with the powers of a Creator:

> Further we must suppose that there is a power, represented by natural selection or the survival of the fittest, always intently watching each slight alteration in the transparent layers, and carefully preserving each which, under varied circumstances, in any way or in any degree, tends to produce a distincter image.[240]

Natural selection...*intently watching*? ...*carefully preserving?* Sounds strangely like deity, doesn't it?

Darwin must surely have been aware of the baggage accompanying any use of the word "Creator," especially in the minds of religious critics reading his newly-published book. How strange, then, Darwin's concluding remarks on the powerful capabilities of natural selection:

> Let this process go on for millions of years; and during each year on millions of individuals of many kinds; and may we not believe that a living optical instrument might thus be formed as superior to one of glass, as the works of the Creator are to those of man?[241]

At face value, Darwin does seem to be referring to "Creator" in the traditional (religious?) sense...all in aid of attempting to prove that natural selection, not "the Creator," is a force quite capable on its own of producing an organ as complex as the human eye. Was Darwin trying to play both sides against the middle so as to placate his religious readers (or his wife), or did he in fact have a lingering appreciation for a divine power existing separately from the stupendous powers of natural selection?

Beyond question, when Darwin was answering his religious critics who claimed that beauty and variety were introduced by the Creator for his own pleasure and the pleasure of man, Darwin uses the term *Creator* in its traditional, religious sense.

> They believe that many structures have been created for the sake of beauty, to the delight of man or the Creator (but this latter point is beyond the scope of scientific discussion), or for the sake of variety, a view already discussed. Such doctrine, if true, would be absolutely fatal to my theory.[242]

A Fail-safe God of the Gaps?

Without chasing a rabbit trail outside the narrow focus of this book, suffice it to say that Darwin does here implicitly acknowledge a duality that he chooses not to further discuss because it wouldn't be "scientific." Yet it is instructive that Darwin also recognized that divine creation (whether for creating beauty for its own sake or otherwise), "if true, would be absolutely fatal to my theory." To bring the discussion back to sex, if Darwin's natural selection cannot explain the phenomenon of sex, the only alternative explanation was one that Darwin was desperate to dismiss. But the harder he tried to dismiss it, the more that alternative hovered over him like a black cloud. Was there really a Creator? Yes (no matter how ill-defined his residual role). Was the Creator the cause of beauty, variety, and distinct species? Definitely not!

Would it be too cynical to suggest that Darwin kept the Creator around in deep reserve as a "God of the Gaps"? Just in case. Just in case natural selection couldn't create something out of nothing. Just in case natural selection couldn't account for the first common ancestors needed for the Grand Theory. Just in case natural selection could not in a billion years (or, even more so, in one indispensable generation!) bridge the gap between mitosis and meiosis. Just in case natural selection could not provide the first set of male and female beings, nor each "first pair" of males and females needed for each of millions of life forms mating and sexually reproducing in unique, often-bizarre ways. And just in case natural selection had any otherwise-fatal flaws.

Whether or not Darwin mentally kept God in his back pocket for emergencies, it's clear that, for Darwin's theory, the "realm of the miraculous" was quite unthinkable. It would mean the certain demise of his overarching hypothesis and the fatal undoing of his Grand Theory. As Peter Brent put it in his definitive biography of Darwin: "The gradual unfolding of species upon species was, he realized, either totally true or totally false. If some miraculous power had the ability to limit it, then its inner logic was destroyed."[243]

Special Creation and a False Dichotomy

Seeing the profuseness of variety and "species" throughout Nature, Darwin had difficulty believing such profuseness could be explained by special creation:

> We can, in short, see why nature is prodigal [extravagant] in variety, though niggard [stingy] in innovation. But why this should be a law of nature if each species has been independently created no man can explain.[244]

If I may be so bold, I believe I can explain it. Darwin painted himself into a corner by wrongly assuming that any *creation* of species by miraculous power necessarily precluded subsequent *variation* of species. On that false premise, Darwin argued that, since variation and a proliferation of "species" is provable beyond all doubt, therefore the independent creation of species could not possibly be true. Were it not for his false premise, Darwin's logic would have been impeccable.

For Darwin and many of his religionist critics, the issue boiled down to the single issue of mutability versus immutability. To whatever extent Darwin's critics were insisting that independently created species were immutable, they were flat wrong. (And desperately wrong to assert that more-recent adaptations were all "special creations"!) But Darwin was in equal error to conclude that the independent creation of species and the mutability of species were mutually exclusive. Unfortunately, neither side seemed to consider that the truth might be somewhere in the middle: not *either/or*, but *both*.

In the closing paragraphs of *The Origin*, Darwin comes as close as ever to conceding the "both" option [with my emphasis]:

> Authors of the highest eminence seem to be fully satisfied with the view that each species has been independently created. To my mind it accords better with what we know of *the laws imposed on matter by the Creator*, that the production

> and extinction of the past and present inhabitants of the world should have been due to secondary causes, like those determining the birth and death of the individual. When I view all beings not as special creations, but as the lineal descendants of some few beings which lived long before the first bed of the Cambrian system was deposited, they seem to me to become ennobled.[245]

Unfortunately, this tossing of a bone to a "Creator" who may have gotten the ball rolling, and may even have set in place certain natural laws, was for Darwin surely more a concession of convenience than an expression of conviction. *Convenience*, since it solved the huge problem of how something could come from nothing. *Not conviction*, since Darwin was convinced that natural selection alone—certainly nothing divinely ordained—explained the true origin of species, all the way from a primordial common progenitor to plants and animals and finally to humankind. Had Darwin wished his audience to believe that there was in fact some divine scheme acting upon or within Nature, he certainly disguised it well.

And here's a pertinent thought for many folks today. Darwin's many references to "the Creator" should be a caution that merely intoning "the Creator" does not preclude virtually emptying the word "Creator" of any real meaning.

At best, Darwin was willing to concede that there might have been (must have been?) a First Cause, as he wrote in his *Autobiography* where he acknowledged:

> ...the extreme difficulty or rather impossibility of conceiving this immense and wonderful universe, including man with his capacity for looking far backwards and far into futurity, as the result of blind chance or necessity. When thus reflecting I feel compelled to look to a First Cause having an intelligent mind in some degree analogous to that of man....[246]

Which brings us back to the Creator as a convenient, theory-saving "God of the Gaps." Or to the blindingly obvious which even the most ardent evolutionists must surely acknowledge.

An Attempt at Dispute Resolution

Is there no way to resolve this dispute between evolutionists and their opponents? I believe there is, and it's not complicated. Let's go back to basics. What do we know for certain? *All living forms are both fixed and evolving—simultaneously. Evolving* in the sense of gradually changing; *fixed* in the sense that there are outer boundaries beyond which gradual changes in form and function cannot happen. That goes first and foremost for the gap between asexual replication and sexual reproduction. There is a fixed, unbridgeable gulf between the two, which no amount of evolution can overcome. The same also goes for each and every living form which generates offspring through radically-different methods of mating and reproduction compared with all other living forms.

What, apart from *a priori* paradigms, precludes the possibility that the earliest progenitors of all subsequent living things appeared suddenly and simultaneously, fully-formed, and capable of either asexual replication or sexual reproduction, depending on their particular, genetically-dictated form and function? If life from non-life is blithely accepted by many if not most evolutionists, or even *something* from *nothing* is virtually assumed in the "Big Bang" hypothesis, why should there be any objection to *this particular version* of life from non-life?

Consider the possibility that every one of the earliest sexually-reproducing progenitors, from the simplest to the most complex, came as a matched set, male and female, and produced offspring having the same basic form and function. Yet consider further that they were capable of variations on theme produced by random mutations, environmental forces, population drift, and the natural selection of fittest forms—at times resulting in varieties ("species"?) unable to interbreed with the original progenitors but equally unable to "step it up" to become other "species" having radically-different sexual equipment and methods of reproducing.

Such a scenario would definitely preclude Darwin's "common ancestor" assumption, but also any notion that specially-created beings were absolutely fixed and immutable from Day One until now. Each position is equally misguided, while the third-way scenario I've suggested avoids the pitfalls of each extreme. More importantly, it avoids both the original "missing sex link" problem and the subsequent "Generation One" problem that evolution's common-ancestor assumption simply can't get around.

If Darwin had not pendulum-swung to an all-or-nothing proposition (either completely *fixed* or completely *evolved*), he easily could have acknowledged without any logical compromise that divine power had set prototype species in place *by species*—even if those species were created with the capacity for evolutionary change by natural selection.

By insisting on an unnecessary extreme, Darwin was forced to ignore any number of factual realities, not the least of which is the persistent, stubborn problem of sex. Unfortunately, the glaring blind spot that Darwin chose to ignore has almost universally been ignored ever since.

As Darwin began Chapter 6 of *The Origin* on "The Difficulties of the Theory," he exuded great confidence in the face of whatever difficulties might be thrown his way:

> Long before the reader has arrived at this part of my work, a crowd of difficulties will have occurred to him. Some of them are so serious that to this day I can hardly reflect on them without being in some degree staggered; but, to the best of my judgment, the greater number are only apparent, and those that are real are not, I think, fatal to the theory.[247]

What's staggering is that, among the many "real" difficulties Darwin addresses, there is not a hint of evolution's sex difficulty—the most obvious problem fatal to his theory. Yet, maybe we shouldn't fault Darwin too much. Who, having formulated what he truly believes to be an elegant idea, would want to acknowledge (perhaps even to himself) a truly fatal flaw threatening to completely spoil such a beautiful theory?

Recap

- The claim that evolutionary sex is the fatal flaw of Darwin's Grand Theory might seem audacious were it not for the fact that it perfectly satisfies Darwin's own test: "If it could be demonstrated that any complex organ existed, which could not possibly have been formed by numerous, successive, slight modifications, my theory would absolutely break down."

- As with most controversies, the truth about evolution lies neither at the extreme of species being completely fixed nor, at the opposite extreme, completely evolved. Rather, living organisms are constantly *evolving* in the sense of gradually changing, yet they remain *fixed* in that there are outer boundaries beyond which gradual changes in form and function (particularly sexual functions) cannot happen, and—fatal to the popular Evolution Story—have never happened.

- Throughout his writings, Darwin occasionally paid lip-service to an Almighty Creator, but in the end convinced himself—and the world—that the faith-prompted notion of a purposeful, designing Creator was as subject to extinction as any more-earthly species along the path of evolution.

- Evolution's sex problem is not a challenge to legitimate science, but—stubbornly ignored and unanswered—is a challenge to scientific objectivity, and to an arbitrary refusal to admit that there might reasonably be a non-materialist explanation about origins which science alone could never explain.

- Rejecting the hopelessly-flawed Evolution Story will only come when the scientific community has the courage to abandon its commitment to philosophical evolutionism in which it has a secure vested interest.

Question: Now that you've thoughtfully considered the premise of this book, do you believe Darwin's Grand Theory can hold together when applying his own test of its validity?

PART FOUR

The Futility of Commingling Evolution and Creation

CHAPTER 11

Cake and Eat It Too?

It is a disgraceful thing for an infidel to hear a Christian, presumably giving the meaning of Holy Scripture, talking nonsense on these topics; and we should take all means to prevent such an embarrassing situation in which people show a vast ignorance in a Christian and laugh it to scorn.

—Augustine

Discussing various theories claiming that God created all of life, specifically including humankind, but that he did so through a process which involved common-origin, microbe-to-man evolution, virtually the same as Darwinism, only without its purely naturalistic assumptions.

If Evolution's vaunted natural selection, acting alone, cannot account for the origin of sex, or for the first male and female pairings of millions of species, might there be another explanation? Many people, both scientists and laymen, think so.

Of course, there are the staunch supporters of divine Creation as set forth in Genesis, whereby God specially created a vast array of fully-mature organisms, both male and female, divinely ordained to reproduce "after their kind." No process. No boundless amounts of biological time. No progression from simple to complex organisms, or from lower to higher species. No common descent. And, most important for the issues raised in this book, no troubling reproduction gaps to worry about. From the very start, sex and gender are fully in place. God simply said, "Let there be sex, and there was sex!"

One has to admit that there is compelling simplicity in the Creation position. Certainly suits the "law of parsimony" and Occam's razor ("Among competing hypotheses, the one with the fewest assumptions should be selected.") When faced with an incomprehensible number of separate, complex assumptions that would have to be made in order to explain, individually and collectively, the origin of light, gravity, the far-flung cosmos, chemical elements, the diversity of biological life, and so on, a Creator God is as simple a hypothesis as one could demand.

And for those who believe in the historical accuracy of the Genesis account, *instantaneous* creation is the most straightforward, reasonable interpretation of the Creation narrative. Of course, given the summary nature of the Genesis account (complete with its opening-chapter overview followed by a close-up reprise of God creating humankind), even many staunch creationists have questioned whether the "six days" are meant to be understood as literal 24-hour days. Others have wondered aloud what we're to understand about the earth being "without form and void." (Does that imply a significant time lag between one divine creative act and others?) Nor are creationists without their own curiosity about a number of intriguing details pertaining to Adam and Eve and their offspring. (Did Cain and Abel marry their sisters, of whom there is no mention?)

But it's certainly fair to say that, *on its face*, there's not the slightest hint of any "progressive creation" from microbe to man. To come up with any kind of gradualistic scenario, you'd have to do some serious reading between the lines.[aa]

[aa] Referring to the Genesis account, including Adam and Eve, skeptic Jerry Coyne states the obvious that "the description in the Bible is straightforward, without the slightest hint that it's an allegory. Now, when Jesus recites parables, like that of the

Surprisingly, a growing chorus of Bible-believers are doing just that. Convinced that microbe-to-man progression is scientific fact, yet also committed to the divine revelation of Scripture, these believers have attempted a hospitable accommodation of what seems at first blush to be two glaringly incongruous viewpoints.

In sophisticated Christian circles, "evolutionary creation" has virtually won the day among Christian scholars. Of particular note is The BioLogos Foundation, which has become the most prestigious and evangelistic proponent of evolutionary creation. Founded by famed geneticist Francis S. Collins, former head of The Genome Project, and now director of the National Institutes of Health, BioLogos is endorsed by such luminaries as Tim Keller, N.T. Wright, Philip Yancey, Os Guinness, Mark Noll, John Ortberg, Richard Mouw, and Andy Crouch (formerly Executive Editor of *Christianity Today* and now with the John Templeton Foundation which maintains close funding ties with BioLogos).

The following BioLogos affirmations are a good summary of "evolutionary creation":

> We believe that God created the universe, the earth, and all life over billions of years.
>
> We believe that the diversity and interrelation of all life on earth are best explained by the God-ordained process of evolution with common descent. Thus, evolution is not in opposition to God, but a means by which God providentially achieves his purposes. Therefore, we reject ideologies that claim that evolution is a purposeless process or that evolution replaces God.
>
> We believe that God created humans in biological continuity with all life on earth, but also as spiritual beings.

Good Samaritan, it's clear that he's simply telling a story to make a point. But that's not the way that Genesis reads." (*Faith vs. Fact*, p. 55).

> God established a unique relationship with humanity by endowing us with his image and calling us to an elevated position within the created order.[248]

Forget the old Creation vs. Evolution debate. "Evolutionary creation" is where it's all happening. Among both Evangelicals and Catholics, it's clearly where "the big dogs" are running these days, not to mention a high percentage of biology professors in Christian colleges and universities.

The Evolution of Evolutionary Creation

For many believers, there's no conscious, articulated amalgamation of Evolution and Creation. The popular Evolution Story of PBS documentaries and National Geographic specials has become so thoroughly entrenched in culture that the average worshiper in the pew assumes Evolution must largely be true. Doesn't everybody know about the dinosaurs, fossils, and billions of years from the "Big Bang"?[ab] On the other hand, even those who are not particularly religious have an intuitive sense that our world is far more complex and orderly than blind evolutionary forces possibly could explain. It takes a certain philosophical obstinacy to reject out of hand what appears for all the world to be intelligent (divine?) design behind it all. (If you have to convince someone of the bloomin' obvious, you can't!)

So, it's not difficult for many Bible-reading, church-going folks simply to assume that Evolution and Creation must somehow comfortably coexist. There's no critical thinking going on here, no heavy-duty analysis, no well-reasoned doctrinal or scientifically-enlightened stance.

[ab] The "Big Bang" and cosmic time-frames—both largely based on logical but faulty extrapolations—are ripe for further discussion outside the bounds of this book. And whatever else might be the history of dinosaurs, like every other sexually-reproducing creature their unique chromosomes, sexual organs, and distinct reproductive methods would have been, from the very first male/female pair, dependent on an evolutionary process incapable of producing the dinosaur sexuality critical to dinosaur existence.

For those who haven't given the subject much thought, the question might be: Why does it matter how God chose to create the universe, whether instantaneously or gradually over billions of years? God created us! What's all the fuss?

Others might be better informed, yet still not be particularly bothered by reason or logic. In a post-modern, multicultural, multi-faith, non-judgmental, subjective-truths generation, acceptance of two mutually-exclusive explanations presents little problem. You say, *both* Darwinian evolution *and* divine Creation? Sure, why not? Whatever!

For the more thoughtful, there can be either conscious or unconscious compartmentalization taking place. Religious folks are not beyond keeping their personal faith siloed from their secular assumptions and actions. Call it a "wall of separation" between the head and the heart. In this category, we might well find serious scientists who fully affirm every tenet of naturalistic Darwinism, yet find comfort in religious experience and expression, including explicit biblical references to our Creator and Maker.

A variation on theme is suggested by Jerry Coyne in his virulently anti-religious book, *Faith vs. Fact*: "For those who want to keep the comforts of their faith but not appear backward or uneducated, there is no choice but to find some rapport between religion and science.... There is no better way to profess modernity than to embellish your theology with science.[249] Intelligent Design proponent William Dembski highlights Coyne's viewpoint this way: "Not to put too fine a point on it, the Darwinian establishment views theistic evolution as a weak-kneed sycophant that desperately wants the respectability that comes with being a full-blooded Darwinist but refuses to follow the logic of Darwinism through to the end."[250]

These informal, unstated accommodations have provided a rich environment for more-reasoned attempts to articulate a formal synthesis. Beginning in Darwin's own day, many have proposed wishful ways in which Evolution could be reconciled with divine Creation. Generally speaking, their various positions are known as "theistic evolution," or in vogue more lately, "evolutionary creation."[ac]

What's Driving This Train?

As we begin to take a closer look, it's important to note the sequence of the arguments. Invariably coming first is an acceptance of the basic Darwinian principle of evolutionary progression. Only thereafter is an attempt made to reconcile that scientifically-based idea with the faith-based Creation narrative in Genesis. The subtitle to Dennis Venema and Scot McKnight's *Adam and the Genome* tells it all: *Reading Scripture after Genetic Science.* "The book has two parts: science and the Bible," says McKnight in his Introduction. "The second part [the Bible] assumes the correctness of the first part [science].[251] (And if that science is *not* correct…?)

You can clearly see the priority of "science first, Scripture second," in the words of Denis Lamoureux (*Evolutionary Creation—A Christian Approach to Evolution*):

> Science has conclusively demonstrated that evolution is the mechanism through which the world was formed, but faith leads to the understanding that it is God's creative method. Therefore, I am expressing my faith when I confidently state that I believe the Father, Son, and Holy Spirit created the universe and life through an ordained, sustained, and design-reflecting evolutionary process.[252]

[ac] Some evolutionary creationists find the term "theistic evolution" unacceptable, believing that the qualifying adjective *theistic* gives primacy to evolution rather than the Creator. Despite such nuanced differences, these labels are sufficiently similar so as to warrant using them interchangeably as we proceed. Any objections to that use are duly noted.

So, science is the locomotive, and faith the caboose. As Lamoureux puts it, "The hermeneutical primacy of science certainly leads to a counterintuitive reading of the Bible."[253] Which is to say that science is the lens through which Scripture must be understood, and if you have to choose between the words of Scripture and science, Scripture must surely be wrong.[254] Says Lamoureux without blinking an eye, "The Bible makes statements about the physical world that are false."[255] Wow.

Other evolutionary creationists would insist that it's our understanding of what the Bible actually says that's the problem. Either way, as William Dembski captures it artfully: "Theistic evolution takes the Darwinian picture of the biological world and baptizes it, identifying this picture with the way God created life."[256] So, it's faith (in evolution) first, followed by a hasty (hermeneutical) baptism.

This order of rationalization is largely true even if most evolutionary creationists began their personal thought processes the other way around. Typically having grown up in Christian homes and churches, they were taught from an early age that God created the heavens and the earth.[ad] When old enough to be taught in school that microbe-to-man evolution is scientific fact, suddenly they were faced with three choices: 1) Reject Evolution as godless propaganda; 2) Reject what their parents and Sunday School teachers taught them, and possibly turn their backs altogether on faith; or 3) Attempt some hopeful rapprochement between Evolution and Creation. Of these three choices, the second is becoming ever more prevalent, with the third gaining ground fast.

Beware the Terminology Game

Anyone thinking critically about "evolutionary creation" should instantly recognize the bait-and-switch going on. Theistic evolutionists typically begin their argument by pointing to all the overwhelming scientific

[ad] Francis Collins is a notable exception, and there may be others, coming to faith from agnosticism or atheism.

evidence supporting the standard model of evolution, then suddenly end up doing a dizzying about-face, insisting that the "evolution" God used as his means of creation was (being divinely orchestrated) modally different from the standard scientific model!

In the hands of theistic evolutionists, the word "evolution" means one thing in one sentence (a wholly naturalistic process accepted by the scientific community), and quite another in the very next sentence (a decidedly supernatural process, which the scientific community resoundingly rejects). "Evolution?" Now you see it, now you don't!

One must take great care here. The word "natural" has a funny way of being all things to all people. Assuming that God supernaturally set in motion everything occurring *naturally* in Nature, then it can rightly be said that the natural process of, say, reproduction and birth, happens as originally ordained by God, but without each and every birth being a supernatural act. For all the times people speak of the birth of a baby as a "miracle," truth is—no matter how mystifyingly amazing—the normal birth process is a natural occurrence. Wondrous, but natural. (By contrast, Jesus' virgin birth and Sarah's conception when past her child-bearing age were clearly *miraculous* events.)

As it happens, this "birth" analogy gives rise to one of the favorite debate-portfolio arguments used by evolutionary creationists. See if you can spot the fallacy....

> ...Christian evolutionists begin by pointing out the remarkable parallels between evolution and human embryological development. They argue that God's action in the creation of each person individually is similar to His activity in the origin of every part of the world collectively.
>
> First, embryological and evolutionary processes are both teleological and ordained by God. At conception, the DNA in a fertilized human egg is fully equipped with the necessary

> information for a person to develop during the nine months of pregnancy....
>
> Second, divine creative action in the origin of individual humans and the entire world is through sustained and continuous natural processes. No Christian believes that while in his or her mother's womb the Lord came out of heaven and dramatically intervened to attach a nose, set an eye, or bore an ear canal. Rather, everyone understands embryological development to be an uninterrupted natural process that God subtly maintains during pregnancy.[257]

Did you catch the fallacy? That which happens in the normal routine of Nature (childbirth) cannot occur until the normal routine itself has been established—which, among other difficulties, requires crossing our now-familiar sex gap. Far from occurring naturally, it would take divine intervention to bridge that gap. This is true whether we're talking about the first-ever simple prototypes of gendered, sexually-reproducing beings, or further up the supposed evolutionary chain in creatures that have wombs capable of so marvelously producing infinitely-complex offspring from a single fertilized egg. Surprisingly, however, theistic evolutionists insist that God never intervened in any of the ways that would be necessary to bridge those gaping gaps. How, then, did those gaps ever get crossed?

There's some delicious irony when evolutionary creationists point to the DNA information contained within the fertilized egg, enabling it to develop gradually and inexorably into an infinitely-complex human being without any special divine intervention during the nine months of its gestation. Crucially, it is the very *lack* of DNA information for sex in any supposed asexual precursor that nullifies any idea that God front-loaded the entire *natural* evolutionary process from the first moment of the "Big Bang." For that very reason, embryonic development is not at all analogous with the originating force that must of necessity have created that wondrous process. Despite all its hope and promise, the cleverly-imagined analogy arrives stillborn.

The End-Game: Chance, Or Certainty?

To be clear, evolutionary creationists are not contending for a deistic God who creates the first living matter, or the "Big Bang," then shuts up shop, neither knowing nor caring where that act of creation eventually would lead. To the contrary, theistic evolutionists insist that God divinely purposed and implemented both the process and the desired outcome.

Why, then, do evolutionary creationists insist on aligning themselves with a process characterized from start to finish by non-directed occurrences with no end-game in mind? The Gospel According to Evolution teaches that, had there been even the slightest course deviation anywhere along the billions of years of evolution, the universe, if it existed at all, would look radically different. And most certainly we would not be having this conversation. Does that sound like the term "evolution" (as in microbe to man) has the slightest capacity to include any purpose, design, or act of God?

Asking rhetorically how God could "be certain of an outcome that included intelligent beings at all," Francis Collins finds assurance that "God could in the moment of creation of the universe know every detail of the future...including...all the biology that led to the formation of life on earth, and the evolution of humans."[258] No, no, no! Wrong answer to a different question. *Foreknowledge* isn't the same as "fore-ordained" or "predetermined." Watching an instant replay of a spectacular touchdown (having already seen the thrilling play that won the game), doesn't mean that the wide-receiver couldn't have dropped the ball. Given divine omniscience, God might just as easily have had "instant *pre*-play" to see a natural unfolding of his divinely-instituted process of evolution that never produced either biological life on earth as we know it or, importantly, humankind.

Pulling out that time-worn trope about the extinction of dinosaurs resulting from an asteroid falling onto the Yucatan peninsula, Collins extends the potential ramifications all the way to us humans. "We probably wouldn't be here," Collins muses, "if that asteroid had not hit Mexico."[259] Really? Human existence is wholly dependent on a rogue asteroid...the outcome of which God merely *foresaw*? Try to envision the outcome if what God foresaw was a natural evolutionary scenario in which we humans never emerged!

According to evolutionary creationists, evolution was definitely a God thing with the specific purpose of bringing us humans into existence. But if all the evolution hype about dinosaur extinction is to be believed, we're just lucky to be here. If that asteroid had not hit Mexico, dinosaurs might still be ruling the earth, and we wouldn't even be a gleam in God's eye. So...which version of "evolution" are we to believe?

Some clarity of thought is needed here. If evolution was *divinely predetermined* so as ultimately to result in humankind—the crowning glory of God's Creation—then it can no longer rightly be called "evolution." By the same token, if evolution is the moment-by-moment outworking of *laws of nature* divinely ordained from the beginning to bring about the human race, any notion of "scientific evolution" is left in shambles. Evolutionary creationists need to do some serious introspection here. Why talk about one thing when you mean something else altogether? Perhaps one could still argue for some kind of progressive creation unassociated with Darwinism, but that would simply be *step-by-step creation* having nothing to do with any accepted definition of "evolution."[ae]

[ae] God, being God, could have chosen to create the universe in some step-by-step fashion. Then again, God being God, could have chosen to create it in the snap of his finger. The issue is not the timing of the overall process but whether God's creative power is being invoked irrationally on behalf of an evolutionary process (considered almost as sacred as God himself) which, in concept and definition, excludes his creative power altogether. The proof of the pudding is that evolutionary creationists never once argue for anything like "step-by-step creation," only the "scientific evolution" generally accepted by the scientific community.

Jerry Coyne puts his finger on the ever-so-subtle subterfuge of redefinition, observing that "Syncretism makes science and religion compatible by redefining one so that it includes the other."[260] It's all the more egregious when there's double-dipping going on. In their effort to redefine Creation to include evolution (which they readily admit), evolutionary creationists end up redefining evolution to include Creation (which they are less eager to acknowledge).

An Unworthy Shell Game

Unfortunately, the slippery use of terminology has turned the typical theistic-evolution explanation into a shell game. Is it truly "evolution" that theistic evolutionists are talking about when they say that God used "evolution" as his means of creation? One thing's for sure: It's not the evolution of Darwinism so highly touted by theistic evolutionists. If, behind the scenes, God is at the controls in any way, shape, form, or fashion, then it is a *supernaturally* conceived and implemented process, not a *natural* process arising and proceeding without external design or action—which is the heart and soul of genuine, textbook evolution.

In the previous chapter, we observed that two radically-different operating systems can be in play simultaneously—as in the coexistence of both a material realm and a separate spiritual realm. Which is to say that there is no inherent conflict in a divinely created world in which Nature (including "bounded evolution") operates pursuant to divinely-ordained laws of nature. By contrast, what cannot logically coexist is an evolutionary process from microbe to man that simultaneously is both directed and non-directed. The "evolution" that theistic evolutionists accept as scientifically valid is the same random, non-directed evolution that appears in virtually all standard biology textbooks. To then turn around and attribute that process of evolution to divine direction is to repudiate the scientifically-accepted evolution upon which the sacred temple of theistic evolution is built.

No matter how appealing its hopeful synthesis might be, "divinely orchestrated evolution"—call it what you will—simply isn't *evolution.* Darwin would have bristled at such an incoherent view of origins. According to Darwin biographer Peter Brent (with my emphasis),

> He resented it when great scientists like Herschel wrote that "an intelligence, guided by purpose, must be continually in action to bias the direction of the steps of change", since *it was precisely the elimination of that hypothetical intelligence that seemed to him his greatest triumph.*[261]

Since by definition from Darwin on down evolution is a wholly *natural* process without some purposeful intelligence involved at any point from microbe to man, it is misleading, disingenuous, and unworthy of intelligent minds to use the term "evolution" when contending that in some supernatural way God was in control of the enterprise.

You say you're not a dogmatic theistic evolutionist, yet nevertheless believe that it's not *impossible* for God to have chosen *some kind of evolutionary process* in rolling out his divine creation? Since "with God all things are possible," one would be wise not to rule out any method at God's disposal. Yet, if the end result of that method—whatever it be—is foreordained, predetermined, and divinely purposed, then that method is not the natural selection of "evolution," acting naturally with no predictable, much less purposed, end-goal. This theoretical amalgamation of the natural and the supernatural is a shotgun wedding attempting a marriage that simply doesn't work.

Both definitionally and conceptually, "evolutionary creation" is an oxymoron of the highest order.

Despite all the rhetoric coming from dogmatic evolutionary creationists about how God wondrously used evolution as his method of creation (how does one put this less than pointedly?), the clandestine, behind-the-scenes role of a Creator God appears to be little more than theological window dressing—merely a cover story for popular consumption. *The truth is, when theistic evolutionists affirm the validity of microbe-to-man evolution, they are doing so without any reference whatsoever to the Genesis account of Creation and its Creator God.* Whatever else they may believe theologically, the battle over the meaning, purpose, and intent of the Genesis account (including its claim that God is the ultimate Creator) is wholly irrelevant to their *academic and intellectual acceptance* of classic, standard, textbook evolution as a valid scientific proposition.

Since that is the case—and this is important—*what the Genesis account has to say one way or the other about a God of Creation is irrelevant to the fact that evolution, as a scientific proposition, is fatally flawed.* It doesn't have to be out-of-step with Genesis to be scientifically suspect. (Or even plausibly be in-step with Genesis by virtue of some idiosyncratic interpretation of Genesis.) The problem is not with *theology,* but *biology. Bible or no Bible, microbe-to-man evolution is simply bad science. Period.*

And there's this to consider as well: If "scientific evolution" is true on its own without any reference to Genesis and a Creator God, then that evolution is wholly naturalistic. If it *needs* God in order to work, then, naturally, it *doesn't work naturally*! And if it doesn't work *naturally* (only *supernaturally*), then it is no longer the "scientific evolution" that theistic evolutionists claim to accept.

If you insist that evolution doesn't *need* God, but simply that, as a matter of divine prerogative, God has been working purposely in and through evolution from beginning to end, you're in no better a position. You've still replaced the standard "scientific evolution" you say you believe in with a designing Creator outside the bounds of science. What's more, that's a Creator you only know explicitly about from a biblical text you insist is theologically compelling but scientifically and factually irrelevant. Try as you might, attempting to play it both ways robs your argument of any logical integrity.

"God of the Gaps"?

Beyond the dodgy logic are the equally-dodgy details. As we've noted repeatedly, evolution's natural selection could not possibly have provided a *mechanism* for bridging between asexual replication and sexual reproduction. The problem only gets worse with evolutionary creation, which has, not just one, but two mechanism problems. The first pertains to evolution's own sex conundrum. If sex is a problem for naturalistic evolution, then it is likewise a problem for whatever purely-naturalistic, non-directed evolution is left over after all the mysterious cameo appearances of God proposed by evolutionary creationists.

The second mechanism problem is explaining *specifically* how God would have worked in, by, or through evolution to achieve his intended goal of human existence. Focusing on evolution's sex problem, in particular, it simply won't do to wishfully employ some subtle, divine presence along the time-line of gradually-unfolding evolution. There are ginormous *gaps* to be jumped or filled!

As we have shown, the first big-time gap is between asexual mitosis and male/female meiosis. Without God's divine hand reaching in to supply a supernatural bridge between the two, evolution would have come to a screeching halt. Then there are the unbridgeable gaps with every first pair of millions of new species. Only a "God of the gaps" intervening divinely and intentionally at each of those millions of points could possibly do the trick.

Yet, if there is a cardinal rule for evolutionary creationists, it's that God never did anything spectacular along evolution's path in order to keep that "natural process" running smoothly. Note Lamoureux's insistence (shared by virtually all evolutionary creationists) that "As the Ordainer and Sustainer of the cosmos, the Creator did not intervene in origins nor does He act dramatically in operations."[262] Francis Collins agrees, including as one of the six fundamental assumptions of theistic evolutionists: "Once evolution got under way, no special supernatural intervention was required."[263]

For evolutionary creationists, of course, their much-touted "built-in bias" was always going on in the background, or perhaps some "fine tuning," and *maybe* a subtle, begrudging mid-course correction here and there, but certainly nothing dramatic.[af] No supernatural dump trucks filling huge gaps, or magnificent suspension bridges floating down from heaven to span any gaping abyss in the process of natural selection.

There's more at stake here than meets the eye. Why are theistic evolutionists so adamant that God didn't swoop down at any point in the process? It's because evolution's signature gradualism must be safeguarded at all cost. Can't possibly risk having God doing anything spectacular, as that would expose the pretense. Everybody would instantly know that the "natural evolutionary process" so highly prized by theistic evolutionists is—in their hands—no longer really *evolution*.

More serious yet, their insistence that God remain in deep background is a powerful reminder of the primacy of Evolution over Creation in the eyes of theistic evolutionists. If ever a civil war broke out between Evolution and Creation where, facing a firing squad (or a doctoral advisor, or tenure committee) you were forced to choose sides, many if not most theistic evolutionists would side more comfortably with Evolution, all the while maintaining a mental reservation that God was somehow "creating" through that process.

[af] At points, Gerald Schroeder (*The Science of God*, xiii) is happy to attribute dramatic corrections to God, saying, "When events veer too far off the desired course, God steps in and redirects the way. Noah's flood is a classic biblical example of God pressing the reset button on society. Was the demise of the dinosaurs…a divine resetting of the earth's ecology?" At other points, Schroeder is more hesitant: "The laws of nature…and the very special conditions on Earth, along with some divine tinkering with those conditions, were quite adequate to orchestrate the flow of the universe toward life" (p. 89). Would "tinkering" include making the leap between mitosis and meiosis? Would "tinkering" be sufficient to provide the first gendered pair of each species?

It's not surprising, then, that you'd be hard pressed to find a single theistic evolutionist who supports some kind of God-of-the-gaps theory—though by some grand irony, theistic evolution is the ultimate God-of-the-gaps theory. From invoking God when needed to explain the origin of the "Big Bang," or to supply Francis Collins's fifteen physical constants without which the universe could not function,[264] or ultimately to endow that quintessential, defining feature of humanity—the Image of God—theistic evolutionists rely on a "God of the gaps" altogether as much as the creationists they love to pin that very charge on.

If scientific evolution was as fool-proof as evolutionary creationists claim, they wouldn't need to enlist God's help in the process. If evolution works at all, it works on its own. So why talk about a Creator God? Truth is, it's not just because there is the ever-looming Genesis account for believers to deal with, but because anyone thinking seriously about the Grand Theory of evolution understands that there are huge gaps begging to be filled. If there weren't so many problematic *gaps*, the term "*God-of-the-gaps*" never would have entered our common lexicon.

Funny how that works out. It used to be the Bible-thumping fundamentalists who were accused of foolishly using God to fill any gaps which science could not explain—an approach which, with each new scientific discovery filling that gap, only served to shrink any need for God. Now, with evolutionary creationists, a God-of-the-gaps approach has reached a whole new level of sophistication. Having God always standing by in the background ("sustaining"), or, better yet, constantly working in the foreground ("maintaining"), there couldn't possibly be any gap, hurdle, or problem of any kind with evolution. (Serendipitously, *both* sides have their own fail-safe theories, since on the evolution side a now-virtually-deified Natural Selection must surely explain whatever gaps, hurdles, or problems might appear. Call it Darwinism-of-the-gaps.)

There's a Bias, That's for Sure!

Taking theistic evolutionists at their word that God doesn't come flying in faster than a speeding bullet, Superman-like, to rescue evolution at any point of peril, the question remains: What actual divine mechanisms do evolutionary creationists ascribe to this Creator God who is using evolution as his creative process of choice?

Not all theistic evolutionists or evolutionary creationists offer the same explanation as to how God has worked in and through evolution. Perhaps the most common view is that God in some omnipotent way *biased* the evolutionary process so as to produce the result he intended all along—including the grand appearance of humankind. Often, that assertion is simply thrown out there without further explanation, as if to say, "Since God undoubtedly created the world, and since evolution undoubtedly happened, the Divine Architect must have drawn up plans that, although appearing at times to be random and unpredictable, were brilliantly designed to move from one level of complexity to another level, and eventually to all levels of life."

Caution is warranted here. It's that other bias—the one that presupposes the validity of non-directed evolution—that creates such dissonance at the point where one attempts to inject an element of divine direction. Directed non-direction isn't a good start. Repurposing an unpurposed process (evolution) in order to prove a purposed process (evolutionary creation) surely isn't something anyone would attempt on purpose, is it? To what possible purpose?

Playing with Dominoes

An evolutionary process having only a vague, built-in bias from the start seems either hopelessly Pollyannaish in its audacious expectations or far too fanciful when it comes to the details. Which leads some evolutionary creationists to take an approach that might well be called the "Divine Domino Theory" whereby an omniscient, designing God sets up all the dominoes in wondrously intricate patterns, then, with his creative finger, pushes over the first domino, setting in motion a process which he never again directly touches. As distinguished from deistic creation, the wondrously intricate patterns thereby produced (including humans) are all intended, purposed, designed, and carefully thought through.

Denis Lamoureux appears to support something very much like the "Divine Domino Theory," saying, "the Creator loaded into the Big Bang the plan and capacity for the universe and life, including humans, to evolve over 10-15 billion years."[265] Sounds like God set it all up and then pushed the first domino, doesn't it? But not even this approach gets us very far. It's one thing to talk about "a plan and capacity," yet quite another to envision the precise mechanism whereby that "capacity" would have ended up exactly according to plan.

For starters, there's all those yawning gaps to deal with—whether it's the paucity of *genuine*, absolutely critical, intermediate forms in the fossil record, or the plethora of missing sex links. Those gaps are not easily bridged solely by some nebulous "plan and capacity" set in motion at the beginning of time. Merely imagine a long line of closely-positioned upright dominoes with occasional gaps ten times the length of a single domino. What might we expect to happen when the rapidly-falling dominoes reach one of those wide gaps? Anyone think that the remaining dominoes will fall over without another push? (Or, are we to assume that the last domino had a built-in divine directive to jump ten times its length to push over the next domino in line? At some point, surely, all these wild scenarios must start making instantaneous creation look good!)

So, the squeeze is on. Evolutionary creationists are caught between a rather feckless "built-in bias" incapable of making dramatic leaps, and the vociferous denials of any periodic, Superman-style divine interventions. When you're being squeezed that hard, you're apt to say almost anything, even if it doesn't make a lot of sense. Merely consider how Lamoureux gets himself all twisted up in a knot:

> ...evolutionary creationists assert that dramatic divine interventions were not employed in the creation of the cosmos and living organisms, including people. Instead, evolution is an unbroken natural process that the Lord sustained throughout eons of time.[266]

Catch the problems? First, there's the same-old, never-ending contradiction of "a *natural* process" that is *divinely guided*. Supernatural naturalism, anyone? Second, is the introduction not merely of an ingenious "plan and capacity loaded into the Big Bang," but now we have a gradually-unfolding process which the Lord God is *actively sustaining* over eons. Surely you don't mean "sustaining" without at some point intervening? (Otherwise, what's the point?) But IF intervening, how is it any longer a *natural* "evolutionary" process? "Cake and eat it too" comes to mind....

When Evolutionary Creation Turns Hyper

The "cake" doesn't become more tempting than when evolutionary creation turns hyper. A good example is found in Richard Beal's little book titled *The Grand Canyon, Evolution, and Intelligent Design*. In describing "providential creation," Beal takes us to a level of divine involvement that borders on radical determinism:

> The first form of life did not come into being accidentally but emerged because God providentially brought together the right molecules under circumstances of his design. The subsequent variations in the progeny of the first protocell and the forces that selected one better adapted cell over another to produce a progression of living forms, were deliberately arranged by God's all-wise and all-powerful activity. God was not a remote superintendent, he was present and

> intimately involved in every atomic detail. Throughout the course of evolutionary history there occurred no slight genetic variation, no modification of the environment, no strike of a predator, no invasion of a parasite, no seemingly trivial event that God did not in his own way control. What we see today, according to providential creationists, is the handiwork of a God who was and is involved minute-by-minute in every circumstance.[267]

At least we're no longer talking about some fuzzy, nebulous "built-in bias"!

Coming from an academic background in zoology and entomology, Beal had a particular fascination with the fossil record found in the Grand Canyon, which Beal found to be compelling evidence of evolutionary progression and common descent. For creationists, it's always the rocks that cause the greatest consternation. If Creation didn't happen gradually, progressively, and upwardly, how does one explain what appears in the various strata of the earth to be the path of evolution?[ag] So, one can sympathize with Beal's effort to put God in charge of every miniscule detail of the evolution history he felt was indelibly written in the rocks.

The biggest problem is the now-familiar logical contradiction which, given Beal's extreme position, becomes exponentially more illogical. If God was actively and intimately involved in *every atomic detail of evolution*, there's not a single shred of "evolution" left. This is no mere quibble over terminology or semantics. Black is not white, and the sun is not the moon!

[ag] Importantly, Beal concedes that "In most instances no rocks are found that contain fossil organisms intermediate between forms in the two strata, as might be expected if evolution took place" (p. 137). And again, "Evidence from the gaps seems compelling enough that both flood geologists and progressive creationists can insist that not all life could possibly have evolved from a single ancestor" (p. 138).

That said, the proof Beal offers in order to demonstrate that purely naturalistic evolution could not possibly explain evolutionary progression is spot on. What's that proof? Evolution's sex problem! As an entomologist, Beal knew his bugs, including bedbugs, whose bizarre sex life begs belief. Says Beal, "To my knowledge no one has ever satisfactorily explained how this most unusual and interesting sexual mechanism could originate through a series of small, successively advantageous steps."[268] Hear, hear!

Of course, what Beal means is that it couldn't have happened in small, incremental steps by any Darwinian understanding, only through Beal's hyper "doctrine of providence" whereby God was 100% involved in *each and every tiny step* of the microbe-to-man process. "In its simplest form, the doctrine of providence states that God controls all events in the universe in such a way that his goals are achieved. God rather than chance or blind fate is in control."[269]

Little wonder that Beal preferred using the term "providential creation" rather than either "theistic evolution" or "evolutionary creation." Despite his many references to "evolution," Beal surely recognized that, by his version of Creation, any conceivable notion of "evolution" is wholly emasculated. Whatever its appropriate name, Beal's singular version is the ultimate in "step-by-step creation." Nothing whatsoever to do with "evolution."

Can We Please Freeze-frame God's Creative Acts?

For all the scholarly talk about reconciling "evolution science" and faith, the conversation invariably pales when put under the microscope of up-close analysis. What do we actually see on the "slide" of evolutionary creation? Virtually nothing of substance. Not even the preeminent exponent of BioLogos, Francis Collins, sheds much light on the details:

> Seeking to populate this otherwise sterile universe with living creatures, God chose the elegant mechanism of evolution to create microbes, plants, and animals of all sorts.[270]

Elegant though the mechanism of evolution may be, what was that mechanism *exactly*? Was it Darwin's elegant *naturalistic* mechanism? If so, how possibly does God's choice of that mechanism not become a supernatural process making complete nonsense of any "natural," non-directed sequence of events? And if not Darwin's elegant mechanism, what *precisely* was God's elegant mechanism?

Our resident skeptic, Coyne, is right about this: "If theistic evolution is to be a truly coherent theory, its proponents must do more than raise it as a theoretical possibility: they must explain what mechanism makes evolution directional, guiding it toward humans, and show us how and where God intervened in that process."[271]

As relayed by Francis Collins, Intelligent Design proponent Michael Behe "has suggested that primitive organisms might have been 'preloaded' with all the genes that would ultimately be necessary for the complex multi-component molecular machines that he considers irreducibly complex. Behe proposes that these sleeping genes were then awakened at an appropriate time, hundreds of millions of years later, when they were needed." [272] To which Collins gives the interesting response: "No primitive organism can be found today that contains this cache of genetic information for future use."[ah] (Behe doesn't actually believe there was a single cell or group of cells that had all the information in their DNA, but possibly, as a tentative view, that the necessary DNA might have been included in the universe or environment as a whole—a proposition which makes "preloaded DNA" even more of an implausible stretch.)

[ah] The burning question for Coyne and all other evolutionists is: Can you point to any exclusively asexual primitive organism that contains a cache of genetic information sufficient for sexual reproduction?

In whatever version, many theistic evolutionists confidently espouse the "preloaded position." So, did God prescribe a time-delayed capsule for his Creation? "Dosage: Take one capsule on Day One (at the Big Bang?), and you're good to go! Make yourself comfy. Sit back, and watch with amazement how it all unfolds." That wishful scenario is as felicitous an explanation as you could ask for, but there's not an iota of credible, verifiable science behind it.

Can We Get Down to Details?

This is as good a point as any to raise a crucial question that every adherent of "evolutionary creation" in any of its forms must answer: What *in fact* would have happened when approaching the abyss at the far edge of an exclusively asexual world? Which of the following options apply?

> A. By an omnisciently-conceived process which God bundled into the first form of life at the "Big Bang," one or more asexual molecules (acting pursuant to a time-delayed divine directive) suddenly split into multiple sets of molecules, instantly acquiring DNA information never before existent in asexual molecules. Over many divinely-directed sequences, they ended up as separate genders, complete with the biological instructions to facilitate a never-before-seen ability to mate, and a method of reproduction (meiosis) never previously set in motion.
>
> B. Acting pursuant to laws of nature omnipotently set in place by God, it was incumbent upon certain asexual organisms to conform their actions to those immutable laws, thereby gradually morphing (by the "law" of natural selection?) into novel forms of sexually-reproducing organisms.
>
> C. With God's divine hand actively guiding each tiny mutation and every environmental change from moment to moment over vast periods of time, asexual molecules

gradually transformed into sexual molecules, eventually being manifested in two distinct genders capable of mating and reproducing in novel ways.

Viewing these options in tight focus, do any of these scenarios make any sense whatsoever? Do any of them come anywhere close to conforming with either Scripture or science? Do they not, rather, demonstrate that attempts to merge such disparate views of origins is not the best of both worlds but indeed the worst of both worlds?

Recap

- If the Evolution Story cannot explain the origin of sex, the traditional Creation Story easily can, wherein God specially created a vast array of organisms, both male and female, divinely ordained to reproduce "after their kind."

- The most straightforward reading of the Genesis account of Creation would suggest that all the first organisms on the planet, whether asexual or sexual, were brought into being "from scratch" by instantaneous creation.

- A growing number of believers—variously known as "theistic evolutionists" or "evolutionary creationists"—accept a less literal interpretation of Genesis, allowing for the belief that God created all living things through the gradual process of microbe-to-man evolution.

- This futile attempt to reconcile the irreconcilable requires undirected direction, unpurposed purpose, supernatural naturalism, and peekaboo definitions of *evolution*—at once extolling the scientific validity of random, unguided, textbook "evolution," yet arguing with equal vigor that God used divinely-guided "evolution" to bring about humankind.

- Any initial appeal of divinely-orchestrated evolution quickly fades when one digs down into the details of what would be required for that dysfunctional marriage of methods to work.

Question: If you believe that a Creator God brought a world full of creatures into existence using the process of natural selection, are you not challenged by the thought that such an inherently undirected, unpurposed process could not randomly evolve so as to produce divinely-purposed human beings, and, more yet, would turn what is affirmed to be a *natural* process (in which God doesn't interfere) into a patently *supernatural* process?

CHAPTER 12

The Soul of the Matter

The reproduction of mankind is a great marvel and mystery. Had God consulted me in the matter, I should have advised him to continue the generation of the species by fashioning them out of clay.

—Martin Luther

Considering the problems which arise when one believes that God used a process of evolution to create humankind. Particularly focusing on the need for God to supernaturally intervene in order to imbue the first human prototypes with a spiritual nature, and, even more so, with a soul fit for eternity.

Given their commitment to scientific evolution, theistic evolutionists are compelled by logical necessity to take the view, (as articulated by BioLogos) that "God created humans in biological continuity with all life on earth"—read: through common descent by the same process of natural selection as all other organisms. Yet given their allegiance to Scripture, evolutionary creationists also cling to the belief that humans are spiritual beings having been endowed by God with his own image. Stand back for this one. Evolutionary apples and Creation oranges are about to be pureed!

As we hear the blender start whirring, it's clear to everyone on all sides that no strictly-evolutionary mechanism possibly could have produced anything remotely spiritual. Richard Dawkins certainly wouldn't take issue with that, nor hard-core creationists. Attempting a precarious middle ground, evolutionary creationists dare to go where angels fear to tread. If evolutionary creationists are to be believed, the Genesis story of God suddenly bringing Adam and Eve into existence from scratch isn't factually true. Rather, humankind gradually emerged from the long and winding process of natural selection and then, *at some point in time*, magically, mystically (divinely!) became spiritually-endowed creatures.

Francis Collins speaks blithely of that process, saying, "Most remarkably, God intentionally chose the same mechanism [evolution] to give rise to special creatures who would have intelligence, a knowledge of right and wrong, free will, and a desire to seek fellowship with him."[273] (Does that include a *soul* that survives death…?[ai])

According to Collins, God chose biological evolution as his means of creating, not just human beings, but *spiritually-endowed* human beings. Hang on! Haven't we already established that no evolutionary mechanism possibly could have produced anything remotely spiritual? So how exactly is this *spiritual* metamorphosis supposed to work? It's not just the origin of sex that's in need of closer scrutiny. Far more important for anyone acknowledging that human beings have a spiritual dimension is the pivotal moment in supposed evolutionary history when God made man "in his image."

Pope John Paul II (echoing his predecessor, Pius XII) offered no bill of particulars on the process, pontificating merely that, "If the origin of the human body comes through living matter which existed previously,

---

[ai] In Scripture, the English word *soul* may mean "the breath of life" (common to all animals who breathe and are thereby animated), or refer to that part of human personhood which (unlike animals) survives death to exist in an afterlife characterized by either reward or punishment. The latter meaning of *soul*, most closely associated with sin and its consequences, is captured in Ezekiel 18: "The soul who sins is the one who will die." Since everyone dies physically, the obvious reference is not simply to physical death of the animated, breathing body, but to the spiritual death of the made-for-eternity soul.

the spiritual soul is created directly by God."[274] *When* "directly"? *How* "directly"? (And was that *living matter* the biblical "dust of the earth," or some hominin precursor?)

C.S. Lewis was equally vague at one point in the evolution of his own thinking about evolution (which must be qualified by his own words and emphasis: "an account of what *may have been* the historical fact"):

> For long centuries, God perfected the animal form which was to become the vehicle of humanity and the image of Himself.... The creature may have existed in this state for ages before it became man.... It was only an animal because all its physical and psychical processes were directed to purely material and natural ends. Then, in the fullness of time, God caused to descend upon this organism, both on its psychology and physiology, a new kind of consciousness...."[275]

Coming from the master of Christian myth and story-telling "In the fullness of time" sounds suspiciously like "Once upon a time..." (In fairness, this comes at a point when Lewis apparently had not been confronted with some of the more problematic details of a scientific premise about which he had fluctuating views. Stay tuned for more Lewisian reflections....)

When you begin to press for details, the questions are endless. Lamoureux, himself, asks two important questions: "When exactly did the ancestors of humans begin to bear God's Image? Five million years ago? Three million years ago? When they began to develop stone tools? At the hunter-gatherer stage? How was divine-likeness given to the precursors of humans?"[276]

Did Lamoureux really intend to say that "divine-likeness" (bearing God's image) was given to *precursors* of humans? Whoa, baby! Is Lamoureux saying that not-yet-fully-evolved precursors of humans were endowed with "divine-likeness," or merely that the moment of endowment was the point at which naturally-evolved precursors became *fully human*?

Either way, when you start heading down this bizarre path, can the details help but get messy?

Speaking of messy, Lamoureux himself asks another fascinating question (with no answer forthcoming). "Was it ordained that humans were to have exactly five fingers? Or did God load flexibility into the Big Bang so that true randomness in Nature would be allowed to evolve different numbers of fingers? Similarly, do the 46 human chromosomes define humanity? Or could the Image of God and sinfulness have been manifested in a creature with a different genetic makeup?"[277] Since he's brought it up, might the image of God have been just as suitable in a three-fingered sloth, or a chimp with 48 chromosomes?

If anyone needs further proof of just how messy it gets, merely consider this mish-mash of the divine and the natural, as Lamoureux waffles his way through a fantastic explanation of how we humans acquired spiritual insight:

> Evolutionary creation contends that the Lord created a 'God center' in our brain through an ordained and sustained evolutionary process. As visual cells evolved in the brain for seeing the physical world, groups of cells emerged during human evolution with the intention to 'see' the spiritual world. In particular, this center is powerfully stimulated by intelligent design in nature. In other words, during evolution God wired our brain for natural revelation."[278]

And you say instantaneous creation is a stretch…? So very interesting, too, how all that "intelligent design in nature" got there by an inherently-blind process of natural selection. The *inherently-blind process* that God himself supposedly chose….

It's hard to know how many evolutionary creationists would agree with Lamoureux's fanciful explanations. The point is that, whatever their particular explanation, the devil for theistic evolutionists (as for all evolutionists) is always in the details. Did God evolve a male first and then a female, or both in tandem? Did these novel, God-endowed humans co-exist with their former not-yet-spiritually-endowed forebears? Whatever happened to those hominin precursors? And assuming a truly gradual evolutionary process, what kind of schizophrenic intermediate creatures would there have been in the transition between non-spiritual hominins and the first spiritually-endowed humans?

If you're looking for unanimity about human origins among evolutionary creationists, you'll be sorely disappointed. Some tell us that God dramatically intervened to implant his image in an Adam-and-Eve-like couple selected from an extant population of evolving pre-humans, whereupon all remaining pre-humans became extinct.

Others say that, at some definitive point, God embedded his image into either all pre-humans existing at the time, or perhaps into a limited group, resulting in a generation of multiple "Adams" and "Eves."[aj] Still others believe (without further explanation) that in some mysterious way God gradually imbued his image through many generations of evolving ancestors.[279] Without question, this gradualistic explanation would most successfully avoid any literal Adam and Eve, but surely that's the least plausible scenario. Half a penis, half a vagina...half a *soul*?

[aj] C.S. Lewis hinted of this position, saying, "We do not know how many of these creatures God made, nor how long they continued in the Paradisal state" (*The Problem of Pain*, 68-71). However, just when you think Lewis didn't believe in the historical Adam of Genesis, we learn from Lewis's biographer, A.N. Wilson, that in a private conversation with Oxford colleague Helen Gardner, Lewis stated that the person from history he would most like to meet in heaven was Adam. When Gardner protested that any "first man" would be an ape-like figure, Lewis responded with disdain: "I see we have a Darwinian in our midst." (A.N. Wilson, *C.S. Lewis: A Biography*, Norton, 1990, 210. See also, John G. West, Ed., *The Magician's Twin*, Discovery Institute Press, 2012, 138.)

A Soul-less Evolution

I say *soul.* Intriguingly, throughout the voluminous pages of theistic-evolution literature you read a lot more about "God's image" being stamped on one or more hapless hominins (reflecting higher consciousness and spirituality) than you do about God creating an eternal *soul.* Not surprising. If evolution can't possibly produce anything spiritual, how much less so a soul destined for existence after death, for which the earthly, physical body is but a temporary vessel? Against all reason, many evolutionary creationists have convinced themselves that the scientific evolution which they are so keen to affirm as God's creative tool could have achieved a "spiritually-conscious" being. But of what possible use would that process be in the formation of an eternal *soul?*[ak]

Maybe our problem here is that we've too often overplayed the *body* and underplayed the *soul.* For anyone acknowledging the teaching of Scripture, we are not physical bodies which just happen to have spiritual souls, but, first and foremost, we are spiritual souls which happen to have physical bodies (and, at the Resurrection, the imperishable "spiritual" bodies Paul whets our imagination to think about[280]). Unlike the animal kingdom, we humans are a fully-interfaced "package deal" of body and soul—never body alone (most especially as merely evolved biological bodies emerging prior to the beginning of human ensoulment), nor ever destined (post-Resurrection) to be disembodied souls, but souls having those spiritual bodies Paul hinted at.

That said, those who profess adherence to Scripture must acknowledge that, in God's economy, the soul is primary, not the body. As in Jesus' rhetorical question: "What good will it be for someone to gain the whole world, yet forfeit their soul? Or what can anyone give in exchange for

---

[ak] Gerald Schroeder (*The Science of God*, 17) defines the human soul as "the *neshama*—instilling free will in humans." Schroeder also speaks of "the creation of the *nefesh*—the soul of animal life—allowing animals choice strongly dictated by instinct and inclination." No mention of a "soul" with an afterlife.

their soul?"[281] And even clearer is Jesus' admonition, "Do not be afraid of those who kill the body but cannot kill the soul."[282] For any believer tempted to conflate biblical doctrine and microbe-to-man evolution, the corollary must surely be: *Don't accept any theory that might remotely explain the body, but never remotely explain the soul.*

An important follow-on question is whether God intentionally shaped the human body to perfectly complement a soul fit for eternity, or was God happy to ensoul just any ol' naturally-evolved physical body? And what if evolution in its blind meanderings had never produced that special, fit-for-a-soul body? If you say, "Well, God made sure it would be the perfect vessel for an eternal soul," you've just abandoned any semblance of *natural* selection, the starting point for theistic evolution.

Then there's this never-mentioned detail: Are evolutionary creationists seriously proposing that the human body into which Christ incarnated was merely the haphazard end-product of an eons-old struggle for survival? For that matter, are believers really saying that a virgin birth of the "second Adam" is perfectly believable, but that the instantaneous creation of the "first Adam" should be dismissed for being scientifically suspect?

What's more, for those professing belief in the Resurrection, Paul's description of the dead rising and those who are still living being transformed from mortal to immortal "in the twinkling of an eye" is cause for pause.[283] Conceivably, Paul's language might whisper of some point in time in which God instantly transformed gradually-evolving hominins into humans having the image of God. But, to the contrary, Paul's words virtually shout that, just as the "new creation" will be instantaneous, God's "first creation" didn't need eons of evolution, mysteriously guided by God's divine hand. It's only a Darwinism strictly limited by blind forces of Nature that has to be accommodated with vast periods of time, not an all-powerful Creator capable of the instantly miraculous. So why on earth would evolutionary creationists "pitch their tents" toward a scientifically-flawed theory that could never produce a soul capable of being resurrected from death or being transformed in an instant to live forever?

Why Use a Shoehorn Unnecessarily?

And consider this. Since Darwin's Grand Theory is fatally flawed by virtue of evolution's sex problem, there's simply no reason for all the Herculean attempts to shoehorn Darwin's spirit-less, soul-devoid origins story into the Genesis creation story. Given evolution's fatal sex problem, any attempt to reconcile Darwinism with divine Creation is totally wasted effort. If *evolution* itself makes no sense, *theistic evolution* makes even less sense.

The same is true of "evolutionary psychology," "evolutionary economics," "evolutionary jurisprudence," "evolutionary morality," and any other "evolutionary" you might wish to think about. If the underlying evolutionary theory is wrong, all the more so whatever particular conclusions are said to follow. And here's a thought to ponder: Would evolutionary creationists be willing to accept the secular-humanist (God-denying, soul-rejecting) perspective about the nature of humankind that in evolutionary psychology is based upon the very evolution science which forms the foundation for evolutionary creation? To protest that the *purely-materialistic, spirit-less* evolution of secularists is not the same *God-ordained, spirit-imbuing* "evolution" they champion is to further expose the doublespeak coming from theistic evolutionists in their effort to reconcile the irreconcilable.

Oh, The Inconsistency of It All

And whatever happened to the cardinal rule of evolutionary creation—that God doesn't dramatically swoop in at any point to interfere with the gradual, inexorably-unfolding evolutionary process through which he, himself, chose to create? What could be more dramatic than intervening miraculously to transform a naturally-evolved hominin into an ensouled spiritual being "made a little lower than the angels"?

Evolutionary creationists need to come clean on this. For all their talk about a natural, evolutionary process leading up to the very first fully-endowed humans, every single one of their explanations inevitably reverts to God breaking through any supposed natural process (no matter how divinely biased) in order to make the extraordinary leap from creatures without spiritual attributes to spiritual beings "made in God's own image."

Beyond that, one can't help but wonder why evolutionary creationists should spill so much ink trying to explain *when* and *how* God made humankind *in his own image*. The first time we ever hear about humans having been made in God's image is in the very same Genesis account dismissed out of hand by evolutionary creationists as not being historical (i.e., factual). So, the obvious question is: If Genesis isn't quite telling us straight-up the real or complete story of human origins, how can we trust that it's right about humankind being created *in the image of God*?

Given the various kinds of literary genre in Scripture, from poetry and prophecy to Job's unique narrative and the apocalyptic language of Revelation, there is scope for understanding the Creation account in Genesis as historical narrative never intended to be taken with wooden literalness. Yet, when Bible-believers get to the point of claiming that Adam was not created from the "dust of the earth" (but, instead, from some precursor hominin) and that Eve was not divinely formed "from Adam's rib" (but, instead, from a long line of evolutionary descent), somebody's seriously lost the plot. Not to mention seriously undermined the whole of Scripture, for which God's sudden, "from scratch" creation of that first couple is the bedrock foundation.

"Oh, what a tangled web we weave...."

Recap

- Both evolutionists and theistic evolutionists agree that natural selection could never in billions of years produce anything spiritual, much less beings bearing the "image of God."
- Having insisted that God never dramatically intervened in the natural evolutionary process from microbe to man, theistic evolutionists do a sudden reversal when God supposedly reached down to imbue prototype hominins with a spiritual nature and even "his own image."
- Evolutionary creationists are all over the board as to the details of when and how God stepped into an otherwise natural process of evolution to supernaturally endow humankind with a spiritual nature.
- According to evolutionary creationists, the first "Adam and Eve" might well have been, not just the one couple featured in Genesis, but perhaps thousands of hominins suddenly or gradually chosen by God to be stamped with his divine image.
- Curiously, evolutionary creationists rarely, if ever, mention *souls* despite the clear teaching of Scripture that spiritual souls, not physical bodies, are the very nature and essence of humanity.

Question: If God's intention from the beginning of Creation was to create souls fit for eternity, wrapped in human flesh for an earthly existence before death, which method of creation do you believe serves that divine purpose better: evolution by natural selection or instantaneous creation?

CHAPTER 13

High View of Science; Low View of Scripture

No man ever believes that the Bible means what it says: He is always convinced that it says what he means.
—George Bernard Shaw

Examining how evolutionary creation compels a reading of Scripture that defies a cohesive, coherent understanding of the Bible and denigrates divine inspiration.

All the right words are said, of course. "The Bible is God's inerrant Word." "The Scriptures are God's revelation to man." Even the phrase "*sola scriptura*" (Scripture alone) is solemnly intoned. The problem is that, for evolutionary creationists, it's never quite *sola* scriptura. Everyone agrees that the Scriptures should be read in their proper context, even if far too often that ideal is honored in the breach. But, for evolutionary creationists, "context" means something altogether different—namely, that the Bible (particularly Genesis) can only be properly understood by going *outside* the biblical text.

According to writers like Scot McKnight (in *Adam and the Genome*), the proper context for the first few chapters of Genesis can only be found in other, Ancient Near Eastern "creation narratives," such as *Enuma Elish*, the *Gilgamesh Epic*, *Atrahasis*, and *The Assur Bilingual Creation.*[284] We're told that the Genesis account of Creation mirrors (or actually borrows from!) the traditions and texts of ancient cultures which focused on "function," not fact; certainly not actual, biological reality. Naturally, it's always Genesis doing the borrowing rather than any possibility that the other Ancient Near Eastern texts might themselves be aberrant, polytheistic variations on theme drawing (as in the "Gossip Game," or "Telephone Game") from an oral tradition first handed down by a historical Adam and his offspring.[al]

Far from honoring *sola scriptura*, still other extrabiblical texts are recruited in aid of the argument that the Apostle Paul, in particular, presents Adam and Eve merely as archetypal literary figures, not historical persons. Enlisted for that purpose are intertestamental writings in the Apocrypha such as *The Wisdom of Solomon* and Ben Sirach's *Ecclesiasticus*, as well as the book of Jubilees (160-150 BC?), the writings of Philo of Alexandria (30 BC-AD 50), Flavius Josephus' first-century history of the Jews, a late first-century apocalypse called *4 Ezra*, and also 2 Baruch (AD 130?). The relevance of extrabiblical texts is problematic enough. How much more so are texts supposedly influencing Paul's thinking but written *after* Paul's epistles.

More curious still, these very texts include references to a real, historical Adam! From *4 Ezra*, for example: "Adam is created by God from dust, and then God breathes into him immortal life."[285] From Philo: Man [Adam] is the "first father of the race" (79); and there's reference to Adam as "first man" and "ancestor of our whole race" (136)."[286] And from the *Wisdom of Solomon*: "Wisdom protected the first-formed

[al] Think also about the many Flood stories in both Ancient New Eastern narratives and other cultures. Considering the disruptive power of flooding, it would not be surprising that flood stories might abound in ancient literature. But given multiple and diverse cross-cultural references to an extraordinary Flood, one would not be amiss in thinking that there must be historical truth to the story of an ancient catastrophic deluge. Once again raising the question: Who first handed down the story? And is there any compelling reason to believe that it didn't originate from eight eyewitnesses whose distant descendants thereafter incorporated that oral history of the Flood into their own, slightly variant folklore?

father of the world, when he alone had been created."[287] Why, then, would McKnight cite these texts in support of the opposite conclusion? Clearly, it's because they also present Adam as being *archetypal*, which McKnight is keen to show is *all* that "literary Adam" was.

McKnight seems not to realize that these quotes expose the false dichotomy at the center of his argument. It's not a case of either/or. Instead, Adam was *both* historical *and* archetypal. That Paul invokes Adam's name as the archetypal sinner in Romans 5, does not exclude a biological, historical Adam. In any number of ways, King David was an archetypal prototype of Christ, yet he was still very much a historical person. Indeed, Jesus—"the second Adam"—was unquestionably the archetypal prototype of righteousness, yet, first and foremost, he was a living, breathing, man-born-of-woman person.

Prima Scriptura?

Unfortunately, evolutionary creationists dishonor not only *sola scriptura*, but also *prima scriptura*, in which Scripture is said to be the ultimate, primary authority. Consider McKnight's subtle legerdemain (presented here with my emphasis): "If we are to read the Bible in context, to let the Bible be *prima scriptura*, and to do so *with our eyes on students of science*, we will need to give far more attention to [various Jewish views of Adam]. For McKnight, keeping "our eyes on students of science" is surely code for giving modern science (i.e., evolution) primacy over the "antiquated science" of Genesis, which ultimately means the primacy of science over Scripture.

We can easily see the primacy of science in the title of Chapter 5: "Four Principles for Reading the Bible after the Human Genome Project." What possible bearing does the Human Genome Project have on how we should read the Bible? Simple. According to genetics, so we are told, the claim that you and I and the rest of humans for all time come from two solitary individuals, Adam and Eve, is "impossible."[288] In other words, the Bible must be read and understood through the lens of modern science. For any "honest, respectful reading of Scripture" (McKnight's oft-repeated mantra), one must acknowledge that science (i.e., evolution) is primary. If someone tells you that humankind descended from a single, biological couple, don't you believe it! That simply can't be true, because science says otherwise. And then begins the whole, long, labored effort to transform what the Bible actually says into something theologically bedazzling, but wholly contrived.[am]

Reading contextually is certainly the right idea...IF you have the right context. For all their insistence on the importance of reading the Bible contextually, evolutionary creationists seem far more interested in a highly speculative context formed by extrabiblical literature than the self-contained context of the Bible itself. Ironically, all this strained use of extrabiblical texts in order to fabricate a hermeneutically-friendly meta-context merely serves to verify one of McKnight's favorite lines: *A text out of context is a pretext.* Never is that more clearly demonstrated than when "context" is in the hands of evolutionary creationists. And why the pretext? No mystery here. It's because the primacy of science over Scripture demands it.

Sticking to the Context of Scripture Itself

If one truly wishes to "respectfully and honestly honor the context of Scripture," it would be a good start to leave one's personal agenda behind when reading a passage like Romans 5:12ff. Reacting to the insistence of many believers that the plan of salvation was made necessary only because a real, historical Adam committed the "original sin" which condemns all of mankind to this day, McKnight argues against the doctrine of original sin in an effort to negate any inference that Paul regarded Adam as a historical person. This, of course, is yet another

---

[am] Even Coyne jumps on the obvious problem, saying, "If you want to read much of the Bible as allegory, you must overturn the history of theology, rewriting it to conform to your liberal, science-friendly faith" (*Faith vs. Fact*, p. 59).

straw-man argument. One need not believe in the doctrine of "original sin" to acknowledge Adam as the archetypal sinner, while each of us remains solely responsible for our own sin ("for all have sinned"). Yet just because he is the archetypal sinner doesn't mean that the "one man," Adam, was merely a "literary Adam" without any real personhood, anymore than that the archetypal righteous one, Christ, was not a real person.

If Adam were *not* a real, historical person, consider how nonsensical verse 14 would be, where Paul says that death (whether physical or spiritual) "reigned from the time of Adam to the time of Moses." What do you think? Is Paul talking about a merely figurative Adam, or a living, breathing Adam with biological descendants? (Was Moses, too, merely figurative?) And while we're in the neighborhood, what do you make of Jude's reference, in verse 14, to "Enoch, the seventh from Adam..."? Sounds like Adam is a real guy with biological offspring, doesn't it? Once again, McKnight has elevated the fallacy of false dichotomy to a new high. But that's not the worst of it....

More serious is McKnight's failure to grapple with the broader context of Paul's complete repertoire of writings. Although he references 1 Corinthians 15 any number of times, McKnight does so only to argue that Paul *uses* Adam to make theological applications. Blinkered in his pursuit of a purely literary Adam, McKnight fails to fully address crucial passages which any honest, respectful analysis of the text would demand. Consider, for example, 1 Corinthians 15:22, "For as in Adam, all die." Wouldn't Adam have to *exist physically* in order to *die physically*? (That obvious logic is verified in the matter-of-fact listing of the genealogical record: "Altogether, Adam lived a total of 930 years, and then he died."[289])[an]

[an] If it's some truly biblical evolution you're craving, you couldn't do better than the longevity of the ancients gradually diminishing to shorter and shorter lifespans. It's not until the deaths of Joseph and Joshua, both at the age of 110, that anything like current life expectancy apparently normalized, after which it became common that folks (like Gideon and David) simply died "at a good old age," which is more in line

And what about 1 Corinthians 15:44-50? "If there is a natural [physical, biological] body, there is also a spiritual body. So it is written: 'The first man Adam became a living being' [obviously the first of all living human beings]…. The first man was of the dust of the earth [neither merely literary, nor from some evolutionary precursor]…. As was the earthly man, so are those who are of the earth… We have borne the image of the earthly man [so if we're like him, Adam must have been like us, a real live historical person]…. Flesh and blood cannot inherit the kingdom of God [suggesting that, like ourselves, Adam was flesh and blood, not merely a literary figure]." If my running commentary in brackets is incorrect, the least McKnight should do is seriously engage with what seems to be an obvious reading.

In other Pauline passages, the historical person of Adam is dispensed with just as cavalierly. With a wave of his hand, McKnight dismisses Paul's reference to Adam being created before Eve (in 1 Timothy 2:11-15) as simply another instance of the "literary Adam" being used for theological purposes.[290] A "literary Adam" is clearly McKnight's story, and he's sticking with it. Presumably that's why McKnight doesn't deign even to mention Paul's discussion in 1 Corinthians 11:8-12. How could the details of who came first, Adam or Eve, be important if we already know (from science) that there couldn't possibly have been any real "first couple" in any event? Dismissed!

Deftly Side-stepping Genealogies

Just as blithely dismissed are the many genealogies in Scripture which point to Adam as the biological progenitor of the race. "The literary Adam of Genesis became the genealogical Adam in the biblical story," McKnight nuances ever so subtly.[291] Which is to say that the genealogies throughout both Old and New Testaments are nothing more than extensions of the (non-historical) Genesis creation myth.

---

with the "three-score-and-ten" reference in Psalm 90:10 ("or eighty if our strength endures"), written, ironically, by Moses, who lived to 120!

Yet how could that possibly be the case when known historical figures populate the lists (with Adam seemingly being the only name in the list not acknowledged as real); when the importance of Jewish identity down through the generations rested on those closely-scrutinized genealogies; when Christ's own lineage is traced back to Adam as proof of his humanity; when list after list is all about "fathers and sons" (even where generations are sometimes omitted); and when other historical details are sometimes included, not to mention the ages of individuals (including Adam) at the time of their deaths? Read for yourself Genesis chapter 5 (from Adam to Moses); 1 Chronicles 1:1-27 (from Adam to Abraham); and Luke 3 (from Jesus to Adam). What do you think? Merely literary, or historically factual?

Perhaps we should be grateful that McKnight doesn't waste our time attempting to put an elaborate spin on these genealogies in the manner of other evolutionary creationists whose tendentious arguments beggar belief. Sure, sometimes there are gaps in the genealogical records; and certainly there are significant differences in the various lists, depending on why a particular genealogy is being presented. But one has to be absolutely blinded by *a priori* assumptions not to take the genealogies at face value as records of real, historical individuals living along the sweep of history.

"Have you considered my servant, Job?"—or at least the Book of Job and its parallels with the Genesis account of creation?[292] Both writings are extraordinary pieces of literature in which Adam and Job are presented as "literary figures." In each story, Satan appears on the scene, also presented in a distinctly literary fashion (raising the interesting question whether that alone makes Satan *merely* literary and therefore not truly real…).

Crucial to our present discussion, Job's "Creation account" is actually far more detailed and cosmic in its sweep than the Genesis record. I challenge you to read Job 38-41 and still believe that God chose to create the universe through a gradual process anything like biological or cosmological evolution. Yet, one has to agree that the style of both biblical accounts is highly literary, and, taken alone, could not easily be argued to rise above the merely literary. Not, that is, until one considers both the clearly historical genealogies presented throughout the breadth of Scripture, and the fact that both the man, Adam, and the man, Job, are referred to as historical persons by other biblical writers.[293]

The point? Even literary genre can be based on literal, historical fact, and it's the genealogies found throughout the Bible that confirm this to be the case with Adam. Evolutionists would be ecstatic to have a fossil record even remotely near as compete and straightforward as the historical record from Adam onward. Whereas evolution's line of common descent is barely visible (indeed, mostly sheer conjecture), the Bible's line of common descent from Adam is in neon lights for all the world to see.

Wouldn't it be nice if, just once, an evolutionary creationist would seriously acknowledge what Paul said to the Athenian philosophers on Mars Hill (Acts 17:22-31)—that "*from one man* he [the God who made the world and everything in it] *made all the nations*, that they should inhabit the whole earth." Does this sound like merely a theological application of a *literary* Adam, or instead a *literal* Adam presented as historical fact to pagan philosophers who would have had very different ideas about human origins? But, of course, mathematical projections based on human genome studies confidently assure us that Paul couldn't possibly be right about that; so, presumably, Paul must have been terribly uninformed. But that's not the worst of it....

Lurching Toward Theological Danger of the Highest Order

McKnight asks a good question: How did Paul know what he knew about Adam? And his answer? "Paul knew about Adam from the Bible.... He did not question this literary-genealogical tradition he inherited from his parents, teachers, and people."[294] So there you have it. Paul was simply repeating what he had always been told about Adam. He believed the myth. He drank the Kool-Aid, and then passed it along to others. (You mean that, when he referred to Adam in Romans 5, Paul actually—if mistakenly—believed that Adam was not just the archetypal sinner but a historical person…?) But not just that, says McKnight. When Paul wrote Romans 1, "it should not then surprise us that many think Paul had read the Wisdom of Solomon and was echoing it there."[295] So, yet more "borrowing" from the extrabiblical lending library….

Is something missing here? Something vital about the nature of Paul's writing? Is there not the small matter of divine inspiration which evolutionary creationists insist they accept? If Paul is to be believed at all, listen carefully as he debunks McKnight's argument: "The gospel I preached is not something that man made up. I did not receive it from any man, nor was I taught it; rather, I received it by revelation from Jesus Christ" (Galatians 1:11-12). Whether that gospel was about Gentile inclusion and the issue of circumcision, or about Adam and the scheme of redemption, the agenda-driven way in which evolutionary creationists manipulate biblical texts to deny whatever is out of step with evolution theory inexorably leads to the most serious of issues: *the inspiration of Scripture.*

If perhaps it's easy to brazenly reinterpret Paul's writings, it's not at all easy to explain away Jesus' own insight into the creation narrative. Jesus of Nazareth would have inherited the same literary-genealogical tradition from Hebrew Scripture as did Paul, so one might argue that Jesus was just perpetuating that tradition when asked about divorce and remarriage (in Matthew 19). Conceivably, Jesus' response, "at the beginning, the Creator made them male and female," might simply be yet another transmission of what was never intended as more than a literary narrative crafted for theological purposes. However, Jesus easily could have made his point about the sanctity and permanence of marriage without referencing the special creation of male and female.

Beyond that, of course, is the altogether marvelous fact that "the Creator" Jesus references is none other than himself—the incarnate Logos by whom all things were created.[296] So, if anyone should know every detail about the origin of humankind, surely it would have been Jesus. Yet, from his lips comes nothing even remotely like McKnight's version of creation, or Lamoureux's, or Collins's, or any other evolutionary creationist.

A Case Study in Overreaching

One cannot help but be surprised and disappointed at the lengths to which even some of our finest theologians are willing to go in making textual wiggle room for the notion of evolutionary creation. As presented on a popular level before a BioLogos audience, for example, the eminent scholar N.T. Wright argues that words from Jesus' own mouth help us to see Creation, not from the usual starting point of an omnipotent, command-issuing Creator of Genesis 1, but rather through the servant-centered, loving, and generous Christ of John 1, by whom and through whom all things were created.

Drawing on Jesus' parable of the seed, Wright analogizes that story to the gradual, random process of evolution: Is not the seed sown prodigally, with both wastage and fruitful production? Does it not develop slowly and secretly, eventually overcoming chaos? Indeed, this deliberative, patient "evolutionary process" is the act of a loving and generous Lord of Creation, in contrast to the arbitrary (unloving?) command of some oriental despot demanding the speedy construction of his palace by an army of architects and builders cowering before him.[297]

To say the least, the "oriental despot" reference is a bizarre, brutish characterization of the traditional view of Creation in which a loving, providential God instantaneously creates the universe from scratch. As for the facile "seed" analogy with evolution, words fail. Can this be serious theology? One shudders to think what the Lord of Creation must think about this brazen misappropriation and exploitation of his Kingdom teaching. As for the evolution Wright champions, he would do well to consider the origin of the sexually-reproducing seed itself. No evolutionary process can possibly explain it, whether Epicurean or non-Epicurean (Wright's distinction between wholly naturalistic evolution and theistic evolution).

Spiritually Serious Consequences

It's time for a serious reality check. Do evolutionary creationists not realize what a dangerous threat to divine inspiration their commitment to evolution has led to? It's not solely a matter of what one sincerely professes or affirms about the inspiration of Scripture, but the serious implications which flow from one's misuse of Scripture. In their minds, the Pharisees could not have revered the Scriptures more highly, yet they were excoriated by Jesus for nullifying those very Scriptures by superimposing their own extra-textual gloss on God's revelation.

And what ultimate irony. Why are evolutionary creationists brought to the brink of blasphemy in diminishing the role of divine inspiration? Somehow, some way—at whatever cost—the biblical account of humankind's common origin in Adam has to be proved wrong in order to safeguard the validity of their sacrosanct theory of common origin in science...even if divine inspiration itself has to scooch over in the pew to make room for the scientifically-enlightened latecomer.

To have any hope of reconciling the irreconcilable, theistic evolutionists have no choice but to make major modifications to two orthodox views. First, there's the wholly naturalistic process of orthodox, "textbook evolution," which somehow has to be transformed into a supernatural, teleological process. Problem is, hard science simply won't allow the luxury of mixing the natural with the supernatural; or, to put it another way, mixing evolutionary oil with creationist water.

The second major modification is no easier: transforming the orthodox Creation story in Genesis into some shape or form that will accommodate evolution science. Here, evolutionary creationists are equally deluded, thinking that, by exclusively theologizing the Creation account, they've neatly resolved the biblical conflict. Adam is dead; long live Adam! But as we've shown, it's not nearly that simple. Delegitimize the Genesis account historically, and you delegitimize the entire Bible theologically.

A Curious Capitulation

May I share a tangential thought as we close this chapter? I'm just curious about trained theologians who espouse and defend theistic evolution. No one doubts their credentials as theologians, but I wonder how many of them have seriously studied the science side of the equation? While I personally know one or two outstanding exceptions, it appears for the most part that theologians pushing evolutionary creation have simply assumed that their esteemed science colleagues must surely be right about evolution. So, instead of asking hard questions about the *science* behind evolution (questions that would be hard for them to ask without some serious digging into the subject), they immediately don their theological caps and begin asking questions about the Creation narrative (spurious, rationalizing questions they would never have been tempted to ask of an otherwise straightforward text if they hadn't already blindly accepted the scientific establishment's Evolution narrative).

You can be sure of one thing: there's no reciprocity going the other direction—where the scientific community defers to classic Creation theology as the starting point for hard questions about the science behind evolution....

When all is said and done, there is one kind of evolution you can truly believe in: *hermeneutical evolution*. It begins with absolute faith in a scientific theory that claims all of life emerged through an upwardly-evolving process of natural selection. By the slightest favorable mutation (merely attributing that process to a supernatural Creator), blind, unguided natural selection takes on direction, design, and purpose. But in order to survive intact, that mutant sterile hybrid must lose an essential bit of its biblical DNA—the historicity of Adam. When the rejection of that crucial DNA is threatened by Paul's multiple references to a historical Adam, his writings undergo an *unnatural selection* process in which clear, contextual meaning mutates into a mishmash of theological manipulation. The stage is then set for an amazing saltational leap in which even the words of the Lord of Creation himself are misappropriated and exploited in aid of the cause, whereupon—at the culmination of this upwardly-evolving process—appears the highest, if subtlest, form of heresy: the undermining of Scripture's divine inspiration.

This is serious stuff we're talking about here. Tugging on Superman's cape and pulling the mask off the old Lone Ranger is nothing compared to messing around with Adam and Eve.[ao] Get that wrong, and a whole long line of unthinkable consequences inexorably follow for people of faith.

Recap

- Despite claiming to honor both *sola scriptura* and *prima scriptura*, evolutionary creationists go to great lengths to de-legitimize the first eleven chapters of Genesis (from Creation to the Flood), saying that those chapters are simply relaying a Jewish variation of other Ancient Near Eastern "creation narratives."

- The problem is not simply that theistic evolutionists consider Adam and Eve merely as archetypal literary figures (or as the first naturally-evolved humans to be "stamped with God's image"), but that an evolutionary view of the Creation story introduces a radically-different hermeneutic in which the

[ao] With apologies to Jim Croce.

reading of Scripture must always be done through the lens of what is accepted by the scientific community.

- Having blithely dismissed the historicity of the Genesis account (including the genealogies which point to a historical, created-from-scratch, Adam and Eve), evolutionary creationists are forced to twist even New Testament passages out of all recognition in order to align with their evolutionary theology.

- The notion of evolutionary creation is so seductive that some of our finest theologians have succumbed to its charms, abandoning textual restraints they would otherwise insist on when doing serious theology.

- No matter how sincerely they pledge allegiance to the inspiration of Scripture, the fact remains that the view taken by theistic evolutionists ultimately challenges, not only the traditional, historical view of Genesis, but the very inspiration of Scripture itself.

Question: If you consider yourself an evolutionary creationist, does it bother you that, in order to accommodate that position, you are virtually forced to read the opening chapters of Genesis in a way that puts an unbearable strain on the rest of Scripture?

CHAPTER 14

Genesis and Genomes

There is something fascinating about science.
One gets such wholesale returns of conjecture out
of such a trifling investment of fact.

—Mark Twain

Engaging with the argument made from human genome analysis that Adam and Eve—traditionally understood to be the specially-created forebears of the human race—could not possibly be the progenitors of the human race.

What are we to think about the stupendous claim made by a number of evolutionary creationists that it is "impossible" for the human race to have descended from Genesis' famed "first couple," since—thanks to the Human Genome Project and its interpretation—we now know for absolute certain that we descended from no fewer than 10,000 hominins-cum-humans some 150,000 years ago? [298] Wow! (And how about that scientifically suspect word *impossible*!)

Who am I to question trained geneticists who've carefully examined the human genome, but since *Adam and the Genome* (co-authored by geneticist Dennis Venema) was written for a general, non-specialist audience, bear with me as I poke around a bit into the assumptions which underlie this extraordinarily bold claim.

The first assumption, of course, is that DNA, the "double helix," and genomes in plants, animals, and humans (not to mention all the other necessary bits and pieces like mRNA, proteins, and amino acids) exist in the first place. The question is, where did all of these critical, stupefyingly-complex features come from? Dare anyone seriously suggest that they self-assembled piecemeal through a non-directed, blind process? (Half a penis, half a vagina, half a "double helix"?[ap]) Venema doesn't come anywhere near broaching that particular question of origins, not even to suggest that God supplied all those wondrous building blocks of life along his "chosen path of evolutionary creation" (which, of course, would blow Venema's cover big-time). This is clearly a no-brainer: Before you interpret data, you must have data.

The second assumption (no surprise) is that life evolved gradually over vast periods of time. For Venema, it's little different from the evolution of language, where languages diverge over time, as, for example, "butter, bread, and green cheese" (in English) and "buter, brea, en griene tsiis" (in West Frisian)—both of which are present-day sequences that are the modified descendants of what was once a common sequence.[299] But the language analogy fails miserably, if for no other reason than that language doesn't have the same built-in (genetic) barriers as is the case with species-exclusive sexual reproduction, a stubborn fact which nixes any process of gradual evolution along a supposed upward path.

The third assumption is that the similarities observed in the genomes of various species obviously suggest common descent with modification. Given that assumption, genomes reflect not just *similar* species, but *related* species. And any observed differences in genomes obviously reflect either "lost features" or "retained features." According to the scientific

---

[ap] You say that it truly was a single helix first, which then doubled? Considering that the structure of even a single helix is so staggeringly complex "no antievolutionists is sufficiently expert to make an intelligent argument about it," how possibly would that first single helix have come into existence gradually by blind forces in Nature? Are we to believe that, according to textbook evolution, divine miracles are out of the question, but evolutionary miracles aren't...?

method of "incomplete lineage sorting" (ILS), "as a population separates, the two new populations will likely both inherit that diversity," but not exactly in the same way, leading to a sometimes-counter-intuitive diversity of species.[300]

Take, for example, the assumed common lineages of gorillas, chimpanzees, and humans. Following an assumed "population separation event," it turns out that in the chimpanzee lineage, the variant "A" was lost, leaving only the variant "a;" whereas in both the human and the gorilla lineages, the variant "a" was lost, leaving only the variant "A."[301] So, despite chimpanzees being our closest relatives on the "species tree," we are said to be closer to gorillas on the "gene tree". The intended take-away is that (obviously!) all three species must have evolved from some precursor population having both the "A" and "a" variants.[aq] Either that, or (dare I even suggest it?) nothing was ever *lost*, but just *is*. Chimpanzees simply *have* the "a" variable, and gorillas and humans both just happen to *have* the "A" variable—nothing to do with inheriting those particular variations from some common precursor species.

Of course, such an alternative explanation would throw a spanner into the works when it comes to calculating the estimated population sizes so key to the proposition that we couldn't possibly have descended from Adam and Eve. "If you have a way to infer what genetic variants were present in a population," says Venema, "you have a way to estimate its

---

[aq] If common descent is valid scientific history, there shouldn't be any differences between placement on the "species tree" based on comparative anatomy and placement on the "gene tree" based on comparisons of DNA, RNA, and proteins. Merely consider the problematic "tree" divergence in the cases of grasses, metazoan animals, lizards and turtles. What's more, the fact that molecular studies notoriously provide widely divergent "gene trees" argues against any clear evidence of common descent. Truth is, whether the "tree" is morphological or molecular, common descent is the invariably assumed starting point, the operating principle always being that similarity indicates inheritance from a common ancestor. With that false assumption, one can build any kind of "tree" you want, even if the branches have to be forcefully grafted in to make it work.

population size."[302] But if there's only *similarity*, not *ancestry* as Venema assumes, then it's "garbage in, garbage out." Without the circularity of evolutionary assumptions, neither population sizes nor the dating of supposed divergent events can be accurately determined.

For evolutionists (including theistic evolutionists), the sophisticated DNA argument is little more than the classic argument from anatomical affinity. Do two species have similar features? Then they must have evolved from a common origin! Do two genomes have sweeping correspondences (or perhaps similar differences in variants)? Then that correspondence must surely mean common descent![ar]

Comparable anatomies, comparable DNA, and comparable genomes, maybe...but, with unique, species-specific sex, any further comparison comes to a screeching halt. Venema himself confirms the obvious (with my emphasis):

> Just as no two speakers speak a language in exactly the same way, no two individuals in a population are genetically identical.... These differences are not a barrier to being part of a species *as long as they do not hinder reproduction with other members of the population.*[303]

That said, Venema then backtracks, parroting the line we've already discussed in detail about what can happen when a single species divides into two groups. In that case, says Venema, "gradual shifts in characteristics can, over long timescales, produce distinct species."[304] "Varieties," of course, would be the more accurate term, since no progression to other, "higher species" having their own genetically-unique mating and reproduction has ever definitively been proved in Nature. Wittingly or not, "variety" is what Venema is really talking about when he says, "Like change within a language, changes within

[ar] Has any thought been given as to how that *similarity itself* came to be? Which is more likely to produce widespread similarity and uniformity: blind evolutionary forces, or some well-thought-out blueprint?

a population are incremental, and every generation remains the 'same species' as their parents and offspring."[305] If all of evolution's incremental changes left both parents and offspring as the "same species," what would account for the millions of *separate* species which evolutionists attribute to that very evolutionary process?

It's not always what's *similar* that tells the story, but what's *dissimilar.* Indeed, when it comes to sex and gender, no amount of anatomical or genetic similarity can account for the abrupt, species-exclusive sexual dissimilarities. Even within a given species, the male and the female could not be more similar either anatomically or genetically, but clearly they don't function the same reproductively. Nor is anyone foolishly suggesting that, simply by virtue of their similarity, the male of each species must surely have evolved from the female, or the other way around (else there could never be a first generation of any species).

What people *are* foolishly suggesting is that modern humans evolved from Neanderthals because we have similar DNA to that which has been discovered in Neanderthal remains. As always, we're right back to the evolutionary assumption that similarity invariably means, not just commonality, but ancestry. *Yet, here's a case where similarity graphically disproves ancestry.* When scientific journals assure us that, along the line of evolutionary descent, there was interbreeding between Neanderthals and "modern humans," what ought to instantly pop into mind? Think especially about the intricacies of "crossing over" and "recombination" in the process of meiosis. There's no room for error here. Mutations aren't tolerated. Chromosomal differences are fatal. To prevent a sterile result, each and every anatomical and internal reproductive process has to be spot-on the same.

Conclusion? Instead of being evolutionary precursors to "modern humans," Neanderthals themselves must have been *fully human.*[as]

[as] Get your genetic history tested at 23andMe, and you'll read these words: "Neanderthals were a group of ancient humans who lived in Europe and Western Asia, and are the closest evolutionary relatives of modern humans." Ancient or

Larger brain boxes and more robust skeletons notwithstanding, if there was mating and reproduction going on, is there any surprise that we would have similar, indeed *shared*, DNA?

To be sure, similarity does indeed suggest "common origin," but what kind of common origin? A chance common origin from precursor species, or a common origin deriving from a common designer? Or to put it another way, are we talking common *descent* or common *denominator*?

Wrong People Asking All the Wrong Questions

Strange, isn't it? Even for theistic evolutionists (some, many, most?), similarity and correspondence couldn't possibly reflect intentional design by a purposeful Creator. You mean, not even if that's the way the Creator wanted his "teleologically-biased method of evolutionary creation" to turn out? Well, no, says Venema. For one thing, the genomes in two similar species are "far more similar to each other than they are functionally required to be."[306] (Is there some law that a Divine Architect is forbidden to use similar design features for similar structures unless functionally required?[at]) And why, Venema asks, would a designer choose to design two genomes "in such a way as to appear to be closely related, especially if your prowess as a designer is such that you can effortlessly design [them] any way you wish?"[307] (Maybe because *that's the way he wished…?*)

And why would the Creator have created everything so similar if he "wanted to convince others that they were separate, independent creations?" (Is it not within the realm of possibility that the Creator might have had other reasons which have nothing whatsoever to do

---

modern, *humans* are still *humans*—all sharing the same human genome with its many genetic variations.

[at] Who's to say that the Creator didn't employ something like Barbara McClintock's famed *transposition*, wherein cells re-arrange chromosomes, pair them in new combinations, and insert them into another part of the genome?

with explaining his actions to mere mortals whom he has created, nor yet trying to convince them of anything?)

Surely, these are all arguments one would expect to hear from *naturalistic evolutionists*, not from *theistic evolutionists*! For those who profess faith in a Creator having the prowess to design human beings any way he might wish, isn't the better question: Why would that omnipotent Creator choose a haphazard, lengthy, and labored process to produce what he intended to produce all along, and could have created from scratch with a mere utterance? Especially since, in order to form creatures with souls fit for eternity, he would have to step in dramatically at some point in any event to do what that haphazard process itself could not possibly do?

I wonder. Does having brilliant insight into the marvelous intricacies of the human genome inspire an unseemly hubris?[au] A familiar word of caution comes to mind: "Surely I spoke of things I did not understand, things too wonderful for me to know."[308]

What About Those 10,000 Hominin Forebears?

No doubt, Venema would say the same of me—that I'm speaking of things I don't understand, things too complicated for me to know. Says Venema, "there does not appear to be anyone in the antievolutionary camp at present with the necessary training to properly understand the evidence, much less offer a compelling case against it."[309] I certainly claim no expertise, but might even the ignorant masses be entitled to ask an impertinent question or two?

For instance, there's that whole set of questions we asked earlier about when and under what circumstances those pre-human hominins became fully human, complete with souls fit for eternity, and bearing the image of God. DNA and genomes are only a small part of the story. Somewhere along the line, mustn't there have been the very first

[au] Might Romans 1:21-23 have some relevance here…?

distinctly *human* couple having all those divinely-gifted attributes? Or is the contention that at some point God specially endowed more than one distinctly hominin couple (perhaps even thousands…) with distinctly human traits, *including their spiritual natures*? Venema's interest seems to be limited to *numbers*, not *natures*; but numbers (and genomes) don't begin to explain natures.

For us uninformed folks, not even the numbers seem to add up. Wouldn't those 10,000 hominins themselves have descended from some *original* hominin couple—the first distinctively hominin pair to evolve from some distinctively non-hominin precursor? So why wouldn't the human genome trace all the way back to *that* First Couple instead of Venema's insistence that "we descend from a population that has never dipped below about 10,000 individuals"?[310] Surely, those 10,000 individuals, themselves, had to come from somewhere![av] And if you say there have never been "distinct species," only gradations along an upward evolutionary path, then doesn't the 10,000 figure—as well as the 150,000-year guestimate—become all the more genetically problematic?[aw]

[av] When you say, "*we* descend from 10,000 individuals," you mean *we humans*? We humans *with souls fit for eternity*? Does that mean we descend from no fewer than 10,000 *ensouled* individuals, or did something extraordinary and miraculous happen between those 10,000 un-ensouled individuals and us? At some point, such incoherence must surely raise serious questions.

[aw] There seems to be no end to the human "lineage," as one study cited by Venema "extended their analysis back approximately 3 million years and found that the population size of our lineage increases the further one goes back in time, with a prior, less severe bottleneck about 500,000 years ago" (p. 53). Yet, why stop there if our lineage goes back to the common origin of all living things billions of years ago? Does our DNA not reflect the entire span of biological evolution? If not, at what point did the DNA genome we humans inherited begin?

And one other impertinent question: If population size increases the further back in time one goes, doesn't that run counter to the standard textbook story of an ever-increasing number of species populations upwardly evolving from fewer and fewer common progenitors the further you go back down the line of evolutionary descent?

Not to mention the *real* conundrum: Go far enough back in time, and there would be no sexual DNA whatsoever, no genome from which one could estimate the size

It goes virtually without saying that the accuracy of calculations depends on the validity of the underlying assumptions. In terms of the human genome, some of those assumptions are at least worth a passing mention. Consider, for example, Venema's claim that "If humans evolved, then we did so as a population. Doesn't everyone know that?"[311] Venema would have us believe that at some point a particular hominin population divided, such that, because of mutations unique to each of the two newly-formed populations, they were no longer able to interbreed and thus became two separate species—presumably the original hominins (now curiously extinct) and us humans.[312]

But before any population can divide, it must first *be* a population! Populations don't just appear out of the blue. Not even the theory of evolution itself assumes that common descent started off with *populations*. (And are we to believe that a population of asexual organisms had a separation event, following which half of them mutated into sexual reproducers!) Like nations, clans, and nuclear families, populations must have a beginning; some running start, some initial progenitors.[ax]

So, first things first. From what did the original hominin population evolve? Venema pooh-poohs the idea that any given species derives from

---

of a sexually-reproducing population of any kind. Where did the first-ever genome come from? And did it not radically change with each emerging species? Uniquely, exclusively? Isn't that why we speak of the *human* genome, as distinct from the unique genomes of chimpanzees and gorillas, even hominins? *Similar* isn't *the same*, or even *common*.

[ax] As everyone also knows, individually we have one-half of our DNA from each of our parents, about 25% from each grandparent, about 12% from our great-grandparents, and so on. If we go back six generations, we've inherited on average something like 1.5% of our DNA from each ancestor. Just curious. How much of our DNA would we have inherited from a population of 10,000 precursors 150,000 years ago? Might this have any bearing on human genome calculations? (And lest anyone be misled by 23andMe's flashy Neanderthal Ancestry reports, if we've inherited only 1.5% of DNA after six generations, can it really be true that one's genome contains, say, 4%-6% of ancient Neanderthal DNA *by way of inheritance*. Surely, it can only mean that 4%-6% of one's individual DNA has correspondence with the DNA found in Neanderthal remains. Again, it's *similarity* or *commonality*, not *ancestry*.)

a single breeding pair, but given what we know of sexual reproduction, how could it be otherwise (except where we're talking about varieties, which are simply variations on theme)? As we have shown, truly distinct species are distinct because they have hard-wired, sexually-unique features so distinctive that no amount of population separation could possibly explain that uniqueness.

And don't forget, it would take millions of improbable "population separations" to explain the millions of different, sexually-unique species. Not to mention how useless this explanation would be in the supposed transition from, say, amphibians to the first-ever reptile. Despite the two classes of species having correspondences in DNA (as, wondrously, does all of Nature), amphibian sex and reptile sex are so definitively distinct that one couldn't possibly evolve gradually from the other. Nor, even less so, could some hypothetical "amphibian population separation" enable a quantum leap to a whole new *population* of never-before-seen reptiles!

As we've previously pointed out, sex isn't like an evolving language where there are no built-in barriers to gradual change. Sex has "pin numbers"! Because of that, we could never have gotten to any hominin population from any population of lower precursor species. For each distinct species, there has to be a first pair! Fully formed; fully compatible; fully unique; fully reproduction-ready from the word "go."

Assume Wings, and Pigs *Can* Fly!

The assumptions just keep coming and coming, including this circular reasoning from Venema: "The baseline expectations *should* be that if humans are the product of an evolutionary process, we arrived at our current state as a population." (And if we didn't evolve...?)

And then there's the "linkage disequilibrium" method of estimating ancestral population size, based on the possible combinations of alleles during the cell divisions that make gametes (eggs and sperm) related to the frequency of "crossing-over"—which first of all assumes the process of fully-gendered *meiosis* that evolution can't begin to explain. Here we go again: If we had any eggs, we could have sperm and eggs, if we had any sperm. Isn't it first things first? Before taking a guess at ancestral populations, don't we need a good explanation as to where all those eggs and sperm came from in the first place?

And what assumptions are at the heart of the "mutation frequency" method? The key to the math is "the human mutation rate and mathematical probability of new mutations spreading in a population or being lost...."[313] Everyone agrees that copying errors (mutations) can supply variety in the DNA, and there seems to be general agreement that a "mutation rate" can be mathematically calculated. Yet the first question must surely be whether *all* DNA differences arose via mutations, or only some percentage of them. In other words, what's your starting point? Naturally, the higher the percentage of assumed mutations, the greater will be the estimates of ancestor populations and time scales. So, as always, we inevitably find ourselves back at square one: Is the genetic variety we observe in the human genome primarily attributable to mutations, or to built-in design (from which there have been subsequent mutations)? Choose your poison: Evolution or Creation?

Even among evolutionists who accept the higher percentage of mutations, there seems to be no consensus regarding outcomes. One study, for example, tells quite a different story: not presumed populations, but single individuals. *Two individuals* in fact. Just two. The headline reads: "Genetic 'Adam & Eve' Chromosome Study Traces All Men To Man Who Lived 135,000 Years Ago."[314] That's *one man* (Y-chromosome Adam), not some large population (nor even a whole population of which that *one man* is a member). And, by the same analysis, a separate (geographically remote!) *one woman* (Mitochondrial Eve) who never mated with the *one man*! (I know. Don't ask. It's a math thing.)

What we're after, of course, are the all-important assumptions on which the "mutation rate" used by each study was based. I'm not sure even

Venema would go for this one: "By assuming a mutation rate anchored to archaeological events (such as the migration of people across the Bering Strait), one team concluded that all males in their global sample shared a single male ancestor in Africa roughly 125,000 to 156,000 years ago."[315] Where, again, did this particular study get its assumed "mutation rate"? From archaeological assumptions. Which merely serves to remind us that assumptions times assumptions equals assumptions squared.

Let's play doctor: Show me your assumptions, and I'll show you mine.

Indeed, show me what part of the human genome you've studied. Isn't it true that humans have ~20,000 protein-coding genes, which is only ~1.5% of DNA in the entire human genome—the 1.5% that common-descent studies are primarily focused on? Is there nothing to be considered in the remaining 98.5%? Nothing that might possibly give reason for pause before making the boldest of proclamations?

And then there's this to think about: "Gene studies always rely on a sample of DNA and, therefore, provide an incomplete picture of human history. For instance, Hammer's group sampled a different group of men than Bustamante's lab did, leading to different estimates of how old common ancestors really are."[316] Certainly, none of these studies come anywhere near proving human lineage from the biblical Adam and Eve, but the fact that two different groups of trained geneticists reach widely-divergent conclusions doesn't inspire confidence in the methods being used. Nor in Venema's assurance that, given the assumptions, "it is straightforward to extrapolate backward from the present into the past."[317] Is that straightforward to a large ancestral population (not fewer than 10,000), or to two, never-mating individuals?

You gotta love those whiz-bang, irrefutable extrapolations! From Darwin's innovative Grand Theory on down, extrapolations invariably present the most fundamental assumption of all—that current observable processes reliably prove merely conjectured processes in the distant past.[ay] When you think about it, such heady extrapolations aren't far removed from *faith*. If only it were a faith we had more scientific reason to believe in....

Adam, or Noah and His Boys?

One might think that Venema and company would capitalize more on the comparison of their human genome calculations with the biblical text. If you take Scripture at face value, all of humanity goes back only indirectly to Adam, descending more directly from Noah's three sons, Shem, Ham, and Japheth. Which must mean that, despite having three times as many progenitors providing a larger gene pool, the time span from Noah's sons to us would be considerably less than from Adam to us, suggesting even stronger support for the human genome argument.[az]

So why not highlight that potentially-favorable time-scale difference? Well, for starters, you wouldn't want to call attention to Genesis 9:19 if you didn't have to—"From these three sons of Noah came all the people who now populate the earth" (meaning, of course, at the time this passage was written; but, by implication, even now as well).[318] That

[ay] Mark Twain penned some choice words about scientific extrapolation: "Now if I wanted to be one of those ponderous scientific people, and 'let on' to prove what had occurred in the remote past by what had occurred in a given time in the recent past...what an opportunity is here!... In the space of one hundred and seventy-six years the Lower Mississippi has shortened itself two hundred and forty-two miles. This is an average of a trifle over one mile and a third per year. Therefore, any calm person, who is not blind or idiotic, can see that in the Old Oolitic Silurian Period, just a million years ago next November, the Lower Mississippi River was upward of one million three hundred thousand miles long, and stuck out over the Gulf of Mexico like a fishing rod." (*Life on the Mississippi*, 1883, pp. 173-176.)

[az] It would be interesting to hear Venema's "take" on the effect of the Flood vis-à-vis the population "bottleneck effect" where there's a severe reduction in population size. "The genetic variation of the new population, then, is dependent on which variants happen to pass through the bottleneck" (p. 47). Even if it might actually strengthen his argument, "the Flood event" likely wasn't a factor in Venema's calculations.

unequivocal biblical assertion certainly puts a premium on any frontal assault such as the human genome calculations.

It's all on the line here. If the Bible is right about this, then the human genome calculations must be off by a mile. Then again, if the Bible is dead wrong on this factual detail, can the Bible be trusted in other, non-scientific matters? Evolutionary creationists make much of the need to reassure vulnerable students that evolution and the Bible are compatible. But can anyone truly believe that this audacious scientific repudiation of plain biblical language will build up the faith of young people who are already entertaining doubts?

That pivotal verse (Genesis 9:19) is "penny ante" compared to the larger problem. Many, if not most, evolutionary creationists reject the historicity not only of Adam but also of Noah and the Flood, presumably including any offspring the "purely literary Noah" might be said to have had. So, it's "in for a penny, in for a pound"—dispensing first with Adam, then with Noah and the Flood. Strange as it may seem for folks who assert long and loud that the Bible is God's inerrant Word, evolutionary creationists are loathe to give historical credence to a catastrophic Flood—first, because they believe that evolution and the geological record refute it, but more especially because, for many creationists, Flood geology and Flood theology are central to their antievolution arguments.[ba]

[ba] In 2 Peter 2:4-9, the Apostle Peter (guided by inspiration?) relied on the historicity of Noah and the Flood as proof of God's certain judgment on the unrighteous and divine rescue of the righteous.

Keep in mind that *historicity* doesn't necessarily mean that descriptive accounts of history must always be "detailed historical accounts" as opposed to "historical narrative" that presents history as story. It just means that, in contrast to fiction or allegory, the underlying facts upon which the more literary narrative is based are historically true. So, when evolutionists (theistic or otherwise) set up the Flood story of Genesis as a straw man argument ("Surely, not all the animals on the earth were in the ark!"), they are arguing against the same kind of historical narrative they would happily accept in any number of other contexts. Merely consider virtually any cinematic depiction of the D-Day Landing, or the Japanese attacking Pearl Harbor.

All of which raises a most interesting question: At what point in time does the factual history of humankind begin to be recorded in Scripture? Was Jesus of Nazareth historical? How about Isaiah, Jeremiah, and the prophets? Saul, David, and Solomon? Moses? Abraham, Isaac, and Jacob? Would we be stretching it to include any part of the line of descent between Abraham and Shem—listed in 1 Chronicles (think about that word *chronicles*!) under the heading in the NIV of "Historical Records From Adam to Abraham"?

A bright line would be helpful. When does the Bible stop being merely literary and become factually historical...?

The Evolution from Exegesis to Eisegesis

One senses that evolutionary creationists are playing the "human genome card" as if that sophisticated scientific process infallibly trumps all else (including Scripture). But in truth there is little new on offer. Wondrous as it is, the human genome is simply data. Data which must be interpreted, like the fossil record, anatomical similarities, and the results of lab tests with fruit flies. The problem is: How are we to interpret the data at hand? If the data are capable of two entirely opposite interpretations, which interpretation should prevail? (Should it be the most parsimonious explanation? It doesn't get more parsimonious than sudden divine creation!) Invariably, this crucial matter of data interpretation becomes the point of departure between evolutionists (including theistic evolutionists) and creationists.

What we see in the interpretation of the human genome is but a microcosm of the larger debate. Consider, for example, this line from Venema, with my own insertions: "Humans do share common ancestors [genetic features] with other apes; apes share common ancestors [genetic

---

Do we dismiss the historicity of either battle simply because it is told in story form? (That said, the problems raised in this book have serious implications for post-Flood speciation, which may raise important questions regarding diluvial scale and scope.)

features] with other mammals; mammals share ancestors [genetic features] with other tetrapod vertebrates...."[319] What is there about common genetic features, other than *a priori* assumptions, that compels evolutionists to conclude that those common features indicate common ancestry rather than common design? Or, in the case of creationists, common design rather than common ancestry?

If we're being honest about it, we're all engaging in more eisegesis than exegesis. *Eisegesis*, of course, is interpreting a text (or data) in such a way that the process introduces one's own presuppositions, agendas, or biases into the text (or data). By contrast, *exegesis* is the exposition of a text (or data) based on a careful, objective analysis "leading out of" the text (or data) itself. In terms of genomes, objective *exegesis* tells us that there are similarities between the genomes of humans and apes. That's all. Full stop. Only by indulging in *eisegesis* can we draw any conclusions about what those similarities (and perhaps dissimilarities) indicate, whether common ancestry or common design.

I confess I don't know all I'd like to know about the human genome, but what I do know (or genuinely believe I know) is that evolutionary creationists use an inordinate amount of eisegesis in interpreting the Scriptures. Absent their evolution-based suppositions, there would be no need for all the polemical gymnastics they employ in attempting to obfuscate the obvious. Which in turn raises my suspicions that their interpretation of scientific data is equally driven by an evolution-based eisegesis. And why do I get the feeling that the process works in reverse as well? That, once you permit yourself the artifice of eisegesis in interpreting scientific data, there's a heady freedom to utilize that same eisegesis in interpreting biblical data.

A Telling Reference to Sex and Gender

Early on in Venema's part of the book, the discussion turns to modern cetaceans (think whales, dolphins, and porpoises) which are said to have descended from terrestrial, tetrapod ancestors. As Venema explains, "Cetacean embryos develop forelimbs *and* hind limbs at the same stage that all mammals do, but later the hind limbs stop developing and regress back into the body wall," hence, "those features...are strongly suggestive that modern cetaceans do indeed descend from terrestrial mammals, even if details remain to be discovered."[320] You mean, minor details like how it would have been remotely possible to evolve *sexually*—species by sexually-exclusive species—from "something *Indohyus*-like, to something Pakicetid-like, and so on through *Ambulocetid*- and Basilosaurid-like forms"? (Have we not moved on from the old "embryonic recall" argument whereby it used to be urged that we humans must have evolved from fish since one might detect in the embryonic stage what looks like fish gills?)

In that context, imagine my surprise to come across the following argument from Venema, which is one of the rare mentions you'll ever find from evolutionary creationists regarding the subject of sex and evolution. If I could highlight the entire quotation in red, I would. Don't skip a word of it!

> It's common for people, upon seeing such evidence for the first time, to begin to reflect on the immense improbability of such large changes taking place repeatedly within a lineage. How could a mutation so large occur to change one animal from one form to another without killing it? How would such an animal breed with anything, unless these rare, massive mutations just happened to occur with a male and female in the same generation?
>
> Isn't this all wildly improbable? Well, yes, such a process would be wildly improbable—so improbable, in fact, that no scientist thinks it could ever happen. This does not pose a problem for evolution, however, because *this is not how evolution works.*[321]

What a concession! What an unintended "admission against interest." Venema has just confirmed the obvious. Think again about Venema's question: *How would an animal breed in a radically different way without a massive change occurring, complete with a male and female in the same generation?* And hear again his answer: *That would be so wildly improbable that no scientist thinks it could happen.* Is this not the very point we've been making in this book—that radical changes in species would require highly complex, simultaneous changes in sexual reproduction that evolution could not possibly provide? Isn't this, in a nutshell, the stubbornly-persistent "Generation One" requirement?

Even if Venema is presenting questions likely to be asked by a skeptic, surely the most uncontroversial part of Venema's statement is that male/female breeding would not occur without "a male and female in the same generation." Female alone? No. Male alone? Not a chance. Sexually-incompatible male and female? No way. Male in one generation, female in the next generation? Are you kidding! Partially-evolved male and female? What do you think!

Then comes the scientifically-acknowledged "wild improbability" of any rare, massive mutation producing any novel pair of prototypes. On that we can all agree. That's not how mutations work. Indeed, as Venema puts it, "this is not how evolution works." How, then, does it work? According to Venema, "Evolution works by incremental change within a population, shifting its average characteristics over long periods of time."[322]

You mean, species-unique penises and vaginas evolving by slow, barely noticeable "incremental changes" within some population of partially-evolved penises and vaginas? If that's the case, how possibly do we ever get to that "same generation" of fully-developed, sexually-distinct male and female of the next higher species? The obvious answer is, we don't! And if evolution's bedrock gradualism is the only way evolution works, then—working as it does *that way*—evolution simply *doesn't work*!

When you say, "evolution doesn't work that way" (meaning that mutations are small and gradual, not massive and immediate), it merely goes to prove that the Grand Theory of Evolution itself doesn't work, whether "that way" or *any way*. It also tells us that, since "unbounded," microbe-to-man evolution is therefore a fatally-flawed explanation for life's origins, there must be some other, non-evolutionary explanation.

Numbers versus *Nature*

So, let's bring it back to the discussion at hand. You say it's *impossible* for the Genesis Creation story to have any historical, factual basis because of what we know from human genome analysis? Here's one even better. Since human genome analysis rests on multiple assumptions flowing from a fatally-flawed theory, would we be terribly off-base to conclude that the highly-touted human-genome argument fails to convince?

What we lesser mortals may not know about the human genome is more than matched by what Venema, McKnight and other theistic evolutionists seem not to appreciate about the countless roadblocks of sexual barriers along the path to the standard theory of evolution. It's back to basics: No fully-gendered sexual reproduction from non-gendered asexual replication, no sexually-reproducing species. No sexually-reproducing species, no common descent. No common descent, no haploids-to-hominins evolution. No haploids-to-hominins evolution, no 10,000 hominins as the minimum number of precursors to our human race. When the assumptions are wrong, the math simply can't be right.

And not just the math, but the science, as well—the science which tells us flat out that real, fixed, genuine species can't possibly interbreed with any other organisms on the face of the earth because they don't have the required (often quite bizarre) mating equipment, compatible chromosomes, species-unique reproductive processes (complete with security systems!), and those all-important distinctive sexual instincts.

The same science which tells us it takes a fully-functioning male and compatible female from the same species to mate and reproduce. The same science which would have to acknowledge that half-a-penis and half-a-vagina would do us no good in producing an ensuing penile/vaginal generation. And the same science which, if it thought about it, would surely concede that Venema's venerated "population hypothesis" couldn't possibly advance the gradualism ball. (Any takers for an *entire population* of half-evolved vaginas…?)

Far more important, when *numbers* can't begin to explain *natures*, alarm bells ought to be going off all around. Don't show us alleles, loci, and codons on the double helix unless you can point us to what part of the human genome explains our being made in the image of God. If it's all about the number of mutating variants in the human population, what do those numbers tell us about the origin of the human soul, which will survive long after it sheds every trace of DNA and inherited genetics? Or is that where a suddenly-intervening Creator swoops in to save the day?

Considering the conflict between *numbers* and *natures*, I wonder if theistic evolutionists have ever stopped to consider what they may have in common with Paul's agonizing struggle (in Romans 7:15-25) between his natural self and his spiritual self. Not unlike Paul, theistic evolutionists seem to be caught in a cognitive dissonance (paralleling Paul's own flesh/spirit dilemma) between the material and the spiritual, the natural and the supernatural. To play off Paul's words, evolutionary creationists run the risk that what they most would like to believe (divine Creation) ends up not being what they truly do believe; and that which they would not like to believe (naturalistic evolution) is what they end up truly believing. To say the least, the fine line they're trying to walk is so precariously thin that falling off on the wrong side would not be unexpected.

Recap

- A popular claim among many evolutionary creationists is that it is impossible for the human race to have descended from Adam and Eve, since human genome analysis conclusively demonstrates that we descended from no fewer than 10,000 hominins some 150,000 years ago.

- Human genome calculations are based on a large number of assumptions, including the assumption that humans evolved from lower species over vast periods of time, that evolution had the ability to provide genomes and the "double helix," that the points of comparison indicate "lost" or "retained" DNA features (which assumes ancestry rather than mere similarity.)

- Comparative DNA suffers from the same fallacy as comparative anatomy, which as likely indicates common design as common descent.

- The validity of human genome studies is effectively negated when trained geneticists reach widely-divergent conclusions (often drawn from focusing on different DNA samples), including the findings of some who believe we evolved from one man and one woman who never mated.

- If the Bible is presenting a story that, on its face, is flatly contradicted by credible science, why should anyone trust the Scriptures in matters not pertaining to science?

Question: Are you tempted to assume that trained geneticists must be right in conclusions even touching on Scripture simply because you don't have the expertise to understand the complexities of the human genome as they do?

CHAPTER 15

Exploring the Subtleties

Is it not equally reasonable to look outside Nature for the real Originator of the natural order?
—C. S. Lewis

Looking at various forms of evolutionary creation put forward by undeclared advocates whose ideas, whether scientific or theological, end up promoting theistic evolution even when only subtly referenced or sometimes overtly denied. Paying special attention to the evolution of C.S. Lewis' thinking about evolution.

Why talk about matters pertaining to theology after studiously avoiding anything other than the crucial biological issues highlighted in this book—namely, evolution's inability to explain the origin of sex? First and foremost, to show that no theological God-talk can overcome evolution's fatal flaw. *Preempting the problem* is one thing (as in creating gender and sexually-reproducing creatures instantaneously from scratch). But when not even God can "fix the problem," that's saying something!

The second reason is to alert believers that wishful efforts to reconcile their faith with a fatally-flawed scientific premise is both futile and foolish. Indeed, wholly unnecessary. If the science behind evolution simply doesn't work, why should that bad science be dragged kicking and screaming into a faith-based alternative?

Then there's simply the need for more critical thinking when reading about origins. Whether coming from science or theology (each is equally as careless as the other), "origins" literature is awash with fuzzy thinking, dodgy logic, and sloppy argumentation. Never is that more evident than when it comes to theistic evolution. Sometimes "cake-and-eat-it-too" is so obvious you can't miss it; at other times, if you're not careful, it sneaks up on you.

Undeclared Theistic Evolutionists

Such is the case, for example, with Michael Denton's *Evolution: Still a Theory in Crisis* (2016). As in his earlier book, *Evolution: A Theory in Crisis* (1985), Denton accepts classical, common-descent evolution, just not Darwin's thesis of gradual adaptation which has been adopted as dogma by most evolutionists. Denton rejects the Darwinian (functionalist) version of adaptive evolution in which the form of organisms is contingent primarily on external environmental forces, leading by sheer happenstance to increasing complexity and apparent order.

Taking a structuralist approach most closely associated with Darwin's contemporary, Richard Owen, Denton argues that the fixed typology of major taxa points to internal, primal patterns embedded in matter and predetermined by natural laws to produce the non-adaptive building blocks upon which all organisms are constructed. Given the scarcity of transitional forms available to bridge the wide gaps between distinct Types, Denton argues that the striking discontinuities along the evolutionary line of common descent are convincing evidence of evolution by saltational leaps (not unlike Stephen Jay Gould's "punctuated equilibrium"). Importantly, however, Denton argues that none of those leaps happened by mere naturalistic chance.

Is he leaving the door slightly ajar for God? Unlike classic theistic evolutionists, Denton never puts forward a case for divine creation, though he seems comfortable with the thought that his structuralist approach suggests design rather than chance. ("As I and others have argued elsewhere, those laws may point to the intelligent design of the universe as uniquely fit for life."[323]) Although the idea of design would be consistent with the intelligent fine-tuning of the cosmos, says Denton, one need not resort to Intelligent Design to accept that his ordered structuralism argument is supported by the greater weight of the evidence over Darwin's haphazard functionalism. Nor does Denton wish to speculate whether the complexity and apparent design of primal patterns necessarily requires a divine Creator, or even an overarching teleology (apparent purpose).

So where does that leave Denton? Certainly, somewhere shy of outright "evolutionary creation," since he never openly affirms that God created the universe using evolution as his creative method. Yet, Denton's insistence on a built-in bias in Nature from the very beginning belies the very teleology Denton is keen to sidestep in order to move on with the non-theological thesis of his book. If not a Creator God, what accounts for that embedded (goal-oriented) bias in living matter; or, more particularly, the laws of nature by which Denton insists that evolution inexorably has proceeded; or indeed the amorphous "nature" which he says "may indeed have 'lent a hand'...," to which the supernatural characteristics of God are necessarily associated? [324]

There are two fundamental problems with Denton's thesis, the first being the glaring lack of any explanation as to how the many gaps between the major phyla possibly could have been bridged. Merely to argue that common descent must have happened in saltational leaps leaves us with nothing but a wishful, blind-faith, science-of-the-gaps explanation. One gets the impression that, having powerfully refuted Darwin's classic gradualism, Denton simply defaults to a saltational approach without feeling any urgent need to explain how evolution could have made those dizzying leaps. The best Denton offers is a persistent resort to "natural law," the crucial source of which is not further elucidated.

Flying the flag for scientific evolution, Denton says [with my emphasis], "I see the Types...and the actualization of their defining homologs during the course of evolution to be the inevitable outcome of *perfectly natural processes*."[325] As a committed evolutionist, it's perfectly natural for him to say so. But listen more closely as Denton agonizingly attempts to square the circle: "This is not to claim that the Types were not actualized by natural processes. I believe they were and that the entire pattern of evolution was prefigured into the order of things from the beginning."[326] *Natural* process? *Entirely prefigured?* With this easily-unnoticed equivocation, Denton runs headlong into the same conundrum as outright theistic evolutionists, hopelessly attempting a bewildering accommodation between the guided and the unguided, the teleological and the non-teleological.

Denton's surely unintended sleight-of-hand requires an audacious redefinition of "natural" to encompass laws of nature (Darwin's own felicitous, omnipresent enabler), which logically necessitates a Lawmaker, thus moving from the purely natural to the supernatural. It's hard to miss the logical inconsistency in Denton's declaration [with my emphasis] that "Typology holds out the immensely beautiful possibility of a *completely scientific* explanation of the phenomenon of life based on *natural law*."[327]

Through what unguided, "completely scientific" natural process could Nature alone possibly have produced its own laws by which to be guided?

Which begs yet another question: Do speeding laws *create* the drivers who speed, or do they not simply encourage already-existent drivers to keep their speed within the posted limits? None other than C.S. Lewis warned against any notion that laws have the ability to create. "In the whole history of the universe the laws of Nature have never produced a single event," says Lewis. "They are the pattern to which every event must conform, provided that it can only be induced to happen."[328]

It becomes all the more curious when Denton suggests that the thesis of his book is consistent with Owen's belief that biology was "pre-ordained into the order of things, part of the fine-tuning of the cosmos for life as manifest on Earth."[329] Who but a Creator *pre-ordains*? Denton's virtual ratification of Owen's term, "pre-ordained," suggests that Denton is an undeclared evolutionary creationist. Or at least a reluctant conscript forced to tacitly acknowledge the necessity of some external force or factor beyond the strictly natural in order to explain where "nature's laws" came from and to account for those amazing saltational leaps.

The second fundamental problem with Denton's argument is his unwavering acceptance of common descent, if not by Darwin's slow-moving adaptive approach (which Denton ably refutes), then by Denton's own gap-leaping, non-adaptive approach. (Says Denton, "descent with modification, can hardly be doubted."[330]) Unfortunately, common descent requires breathtaking leaps that none of Denton's embedded homologs possibly could have achieved, among the most important being the origin of sex.

Compounding Denton's problem with evolutionary sex is the inability of evolution (either his version or Darwin's) to provide the first-ever male and sexually compatible female of each species, with just the right matching number of chromosomes, unique sexual organs, and exclusive reproduction processes. Even within major taxa or clades where common primal patterns are clearly evident, not one of those related species can have sex and reproduce with any other form of that Type. Nor can a taxa-wide homolog begin to explain the origin of separate, but compatible genders. In short, structuralism alone can't explain the simultaneous appearance of reproductive-ready males and females in the crucial go-or-no-go first generation of each species that is so indispensable to common descent.

Even conveniently assuming all the felicitous laws of nature you wish, it nevertheless remains that Denton's saltation thesis simply can't provide Fed-X-like, on-time delivery for the Baroquely-diffuse world of sexual reproduction. And if you're not able to produce the sex so ubiquitous in Nature, then you can forget common descent. More important yet, where there's no *actual, factual* common descent, it's all the more senseless to talk about *prefigured* or *pre-ordained* common descent.

Theistic Evolution Hidden Behind a Pretense of Neutrality

So far, our discussion has focused on various versions of evolutionary creation where the process of evolution is either explicitly or implicitly argued to be God's chosen method of divine creation. The more explicit versions typically accept that the Genesis account is sufficiently pliable so as to accommodate a divinely initiated and guided form of scientific evolution. The debate thus centers on which method of creation Genesis speaks to, whether the traditional view of instantaneous creation or the more scientifically-based view of some gradualistic, evolutionary creation.

In his widely-read book, *The Lost World of Genesis One*, Wheaton University's John Walton (a member of BioLogos' Advisory Council) posits a novel, third way of understanding the Genesis account which claims not to champion either of the two usual interpretations, but rather to de-materialize the Creation story altogether. Walton's thesis, which he calls the "cosmic temple inauguration" view, is all about theology, having nothing to do with "creation" at all. Rather, the *seven days* describe the inauguration of the cosmos whereby "the cosmos is being given its functions as God's temple, where God takes up his residence and from where he runs the cosmos." Accordingly, "on day one God created the basis for time; day two the basis for weather; and day three the basis for food," and so on.[331] The key word is *function*, not *creation*. "As an account of functional origins, it offers no clear information about material origins."[332]

Walton's view generally assumes that, prior to the seven-day "temple inauguration," the initial stages of the material universe had already taken place, whenever and by whatever method God may have chosen. Since "creation" details are not relevant to the point being made by the writer of Genesis, says Walton, it's no use arguing one way or the other about how God may have implemented his creation. All we know, and need to know, is that God did it! So, let's move on to the wonderful spiritual insights to be gleaned from this magnificent theological story. If only....

So many objections, so little space. Question One is: How in the world does a Bible scholar offer this idiosyncratic interpretation with a straight face? Walton's answer is that centuries of readers have missed the true, intended message of Genesis for failure to understand "the lost world" of Genesis. "As in the rest of the ancient world, the Israelites were much more attuned to the functions of the cosmos than to the material of the cosmos. The functions of the world were more important to them and more interesting to them. They had little concern for the material structures; significance lay in who was in charge and made it work."[333]

So, we're not to take the Genesis account literally? Oh, no, Walton insists, absolutely literally! The "temple inauguration" view *is* the literal meaning of the text! "The reading this book proposes is precisely what the Genesis author and audience would have understood."[334] And how are we supposed to know that? Apparently, if we were all literate in Hebrew and its cultural context, it would be crystal clear. "Truly learning the language requires leaving English behind, entering the world of the text and understanding the language in its Hebrew context without creating English words."[335]

Makes sense, I suppose, but is it not curious that none of the biblical writers (nor Jesus himself)—all of whom knew Hebrew and Hebrew culture intimately—seem to have shared Walton's view of a de-materialized Creation story? Is there the slightest hint of Walton's "temple inauguration" view in any passage, Old Testament or New? Surely, whatever might be a "lost world" to us, would not have been lost on them.

And if the Genesis account has nothing to do with the creation of anything material, why the sudden screeching of tires when it comes to the creation of Adam and Eve?[bb]

> The image of God and the sinful act of disobedience dooming all of humanity are biblical and theological realities linking us to Adam and Eve, whom the biblical text treats as historic individuals (as indicated by their role in genealogies). That God is the Creator of human beings must be taken seriously.
>
> Whatever evolutionary processes led to the development of animal life, primates and even prehuman hominids, my theological convictions lead me to posit substantive discontinuity between that process and the creation of the historical Adam and Eve.[336]

Note, first of all, how Walton tips his hand here, assuming a process of evolution by common descent leading up to God's creation of the first population of human beings—which instantly raises evolution's insurmountable problem with sex. If we're talking about God directly creating Adam and Eve, it's easy enough to point to their being specifically made male and female (or functionally were "designated male and female," as Walton puts it oddly[337]). But it's by no means easy

---

[bb] On the validity of using biblical genealogies to prove a historical Adam, Walton breaks with evolutionary creationists, such as Lamoureux, who offer a portfolio of labored arguments that the genealogies do not, in fact, support a historical Adam. He is also adrift from McKnight's argument that there was no historical Adam, only a "literary Adam."

In *Adam and the Genome* (p. 109), McKnight is less than straightforward when he co-opts Walton's archetypal interpretation of 1 Corinthians 15, while breezing past a line (with my emphasis) from Walton's discussion in *Four Views of The Historical Adam* (p. 107): "It is insufficient to bring in biology simply because *Christ was biologically descended from Adam.*" While McKnight is happy to enlist Walton's support for an archetypal reading of Paul's argument, he (unlike Walton) doesn't believe that Christ was biologically descended from Adam—once again exposing McKnight's cherry-picking, false-dichotomy approach to the text, whether it be Paul's or Walton's.

to explain the development of gender and sex in the process of natural selection that Walton blithely assumes in the evolutionary build-up to Adam and Eve's existence (the two being specially drawn from a larger, extant human population imbued with "the image of God").

From there, the problems only multiply. When God gave plants and animals the *function* of being fruitful and multiplying, says Walton, "God created them capable of doing so."[338] That obviously works if we're talking about instantaneous Creation, but doesn't fly at all if we're talking about natural selection. Since Walton assigns all pre-human "creation" to a process of "scientific evolution" (continuously fudged by insisting God had his fingerprints all over the process), either Walton has to deal with evolution's fatal flaw or resort to a God-of-the gaps making the millions of sex and gender leaps that would have been necessary.

And then there's that ultimate leap. By conceding God's abrupt intervention into the evolutionary process in order to stamp the first "humans" with his divine image, Walton once again falls back on a God-of-the-gaps approach…based solely on his theological convictions. This, despite conflicting with the prevailing scientific consensus that Walton proposes we ought to accept as valid until proven otherwise.

And where does Walton get his theological convictions that Adam and Eve were fully-material historical human beings? Where else but from Genesis? Behold, that which once was lost is now found!

In politics, we've grown accustomed to "non-denial denials" whereby one can carefully nuance what looks for all the world like a denial in such a way as to maintain plausible deniability. Walton provides us another example when he says: "It should be noted that this book is *not* promoting evolution."[339] One could be excused for thinking otherwise. At the very least, we hear Walton saying: "Biological evolution is not the enemy of the Bible and theology."[340]

Of one thing you can be sure: Walton is *not* promoting instantaneous divine creation. Rather, from cover to cover we see Walton permitting, enabling, and promoting "scientific evolution" in the form of evolutionary creation.

> Though the Bible upholds the idea that God is *responsible* for all origins (functional, material or otherwise), if the Bible does not offer an *account* of material origins we are free to consider contemporary explanations of origins on their own merits, as long as God is seen as ultimately responsible. Therefore whatever explanation scientists may offer in their attempts to explain origins, we could theoretically adopt it as a description of God's handiwork.... If there was a big bang, that is a description of how God's creation work was accomplished.... If various life forms developed over time, God is at work.[341]

So, how are we to know what method God used to create the universe? Ask the scientific community—the same scientific community whose naturalistic, materialistic explanation leaves God completely out of the picture. [bc] And how's this for air-tight logic? "Science cannot offer an unbiblical view of material origins, because there is no biblical view of material origins aside from the very general idea that whatever happened, whenever it happened, and however it happened, God did it."[342] Is Walton saying it wouldn't be unbiblical even if the prevailing view of the scientific community is that all of life, including humans, evolved through a process totally devoid of God's input? Well...no, not that...because God was definitely behind the process, whatever it was!

[bc] Walton's argument begins with a false premise leading to a false conclusion: [False Premise] Since "the Bible does not offer an account of material origins," then [False Conclusion] evolution is legitimately in play. The opposite implication is worth considering. What is the logical conclusion if Genesis does, in fact, offer an account of material origins that makes no reference to any gradualistic, evolutionary process, but rather appears on its face to be instantaneous? Does this not explain the desperate attempts to twist the Genesis text out of all reasonable recognition?

Walton's Fairytale View of Evolution and Education

Walton's apparent goal is to be a neutral force between hard-core creationists and strident evolutionists, creating separate and distinct jurisdictions for Scripture and science. As it happens, Walton has taken a page from the Supreme Court in matters between Church and State. Having declared judicial neutrality, the Court consistently rules in favor of secularism as against faith (best seen in the battles over evolution!). By default, "judicial neutrality" invariably evicts faith from the public square, inexorably becoming anti-religious.[bd]

The same pernicious "neutrality" factor is at work in Walton's book. Thanks to Walton's imaginative interpretation of Genesis, being neutral means permitting a forum only for Evolution, not Creation. Dismissing Creation as definitionally unscientific, Walton urges that the scientific consensus regarding origins (standard, textbook evolution) be taught in public schools…only without either teleology (purpose and design) or dysteleogy (no discernable purpose). "It must be teleologically neutral."[343] *Teleologically neutral?* Do you not know? Hath it not been heard? Just as water IS wet, and accidents ARE purposeless, evolution IS dysteleological. Which only makes a mockery of Walton's preference for the oxymoronic term "teleological evolution."[344] Whether definitionally or conceptually, there's no such animal.

Left to Walton, the only side of the story that can be told to children in public schools is that we humans evolved through a wholly natural, undirected, blind, and meaningless process. There must not even be a stage-whisper about God being behind that process (a position which Walton personally asserts to be true), nor especially about God's divine intervention to purposely create human beings (as Walton believes to be factually and historically true). If public education is the only input

---

[bd] One can't help but think of Stephen Jay Gould's "nonoverlapping magisterium," which ought to be an immediate caution. In the hands of both evolutionists and theistic evolutionists, the magisterium of science not only *overlaps* the magisterium of religion but arrogantly and brutishly *overpowers* it.

on origins that most students will ever get, we've just raised multiple generations of secularists for whom life has no inherent meaning.

So, what should we do about Evolution's sex problem? Can it even be discussed? Walton naïvely proposes that public schools should teach both the strengths and weaknesses of evolution.[345] Really? Does anybody seriously think that the scientific community would stand idly by and allow an open and honest discussion of Evolution's theory-busting sex problem? If in fact there is a fatal flaw in Darwinism, what's the only reasonable alternative to take its place? Not only would that alternative undermine the secular worldview of evolutionists, but it would automatically bring God back into the classroom, breaching the now-sacred "wall of separation" between Church and State. (You can almost hear evolutionists screaming that challenging Evolution on its most fundamental facts is a sinister ploy, as with Intelligent Design, to introduce the idea of divine Creation.)

Compounding the confusion, Walton's premise would say that students in private religious schools (even Sunday school?) should be taught that Genesis has nothing whatsoever to say about how God created the universe…and therefore, by logical extension, that they should accept the prevailing scientific (God-void) explanation. So, how do believers ever get God into the picture for schoolchildren? Well, of course, Genesis teaches…!

All this coming from a Christian educator whose book is being read by students in Christian universities across the nation. Is Walton's approach seriously going to prevent young people from abandoning their faith when Evolution and the Bible collide? Are they really so gullible?

Whether we're talking hard-core evolutionary creation like that of Collins, Lamoureux, Venema, and McKnight, or the soft and subtle versions implicit in Denton's side-bar with Darwinism or Walton's tortured interpretation of Genesis, the truth is that trying to reconcile the irreconcilable is fool's play. It certainly doesn't rescue Evolution's insurmountable sex problem, nor, in the larger frame, Darwin's Grand Theory which stubbornly defies any attempt to transform its purely natural process into something supernatural. Far from bolstering a flawed scientific theory, this facile attempt at a blended synthesis merely serves to diminish the very faith commitment evolutionary creationists wish to honor.

How Big Is Your God?

What would real faith in a divine Creator look like? If God has the divine ability to direct evolution toward the ultimate goal of creating humankind (as evolutionary creationists argue), he must surely have the divine ability to instantaneously create millions of creatures having strikingly similar genomes. Not to mention creating a full spectrum of species from "high" to "low" whose comparative DNA could readily be charted on a sliding scale. And not as some devious, divine plot to *fool* us (a tired straw-man argument), but simply because he chose to do it that way. (Isn't it the evolutionary creationists who insist that God could have chosen any shape and form of creation he desired…?)

Quite incredibly, a truly perverse objection is forthcoming from some evolutionary creationists. Dramatically creating millions of species from scratch, they argue, reflects a low view of God, because instantaneously speaking millions of unique species into existence would be, well, *too easy* for God! (Yes, you read that right.) Compared to such a humdrum miracle as instantaneous creation, creating humankind and the whole of Nature by a far more circuitous process of natural selection and survival-of-the-fittest is the coolest form of design possible. How so? Quoting verbatim from private correspondence with a bright, well-informed theistic evolutionist, "It's immensely more difficult to start with a single cell and end up with zebras eating grass than to create the grass and then create the zebras in a presto miraculous moment." What can one say? That's some serious Kool-Aid.

So, we are told, divinely-ordained evolution actually reflects a far higher view of God than does instantaneous creation. Ah...*a higher view of God.* And just that quickly we are transported to a screaming schoolyard and the childish taunt: "*My* God's bigger than *your* God!" And just like that, the matter is settled. The kid with the bigger God wins!

When it comes to how big God is, there's no place for schoolyard bullies. God is big enough to confound the wise, no matter how he chose to create us. But the fact remains that, the bigger your God is in the process of creation, the less anything remotely like Darwinian evolution is in play. And the greater the role genuine evolution plays, the less your "Big God" has anything to do with it. (Lest we forget, Darwin made the Evolution Story he sold to the world so big that there was no room for a God of any size.)

The "bigger God argument" is as offensive as it is lame, and even more hypocritical. Were someone to chart on a sliding scale the comparative "faith DNA" of theistic evolutionists in either God or Darwin, the results would hardly be encouraging for God. God's name may be prominently displayed over the front door (in BIG letters), but it's at the back door, marked "Darwin," where everybody's going in. Whatever their personal religious faith (no matter how strong and sincere), when the talk turns to science, an emaciated, feeble faith in divine Creation appears.

Fatally-flawed Theory; Fatally-flawed Philosophy

In the larger frame, *religious theists* who so eagerly line up with *secular evolutionists* are unwittingly lending credence, not simply to a fatally-flawed scientific theory, but to the secular creed most closely associated with evolution theory. The swelling number of evolutionary creationists is testimony that one doesn't have to be an atheist to believe in evolution. But it's hard to ignore the strikingly high percentage of hard-core evolutionists who happen to be atheists and hard-core atheists who happen to be evolutionists. Evolution is more than a scientific theory. As an alternative explanation of the meaning of life (that life has no intrinsic meaning!), Evolution has become the religion of the non-religious. The faith of the faithless.

Theistic evolutionists would do well to step back and consider that they are not natural allies with faithless "evolutionism" and its inherently amoral outworkings. *Why* are they not natural allies? What is there about this particular scientific theory that so often leads its proponents away from faith? Surely, it can't be so difficult to see. Unlike other scientific theories, Evolution—*as a theory*—dares to claim it doesn't *need* God nor *heed* God. In the strictest sense of the word, Evolution is by its very nature *godless.*

For all their good intentions (and often scientific brilliance), evolutionary creationists are hopelessly naïve. Any attempt to coax a theory wholly devoid of God into mystically mutating into a God-created, God-infused, God-directed, and God-controlled process can only result in a mutation that will never be selected for survival. Indeed (in terms that Francis Collins and his Human Genome Project colleagues would understand), its damaging "misspelling" of the "language of God" in the genome of life virtually assures that it can only produce nonsensical gibberish.

In the Valley of Decision

If we're all being honest, no matter in which direction we look there are enough questions to challenge any serious thinker. Little wonder that so profound a thinker and writer as C.S. Lewis seemed at once drawn by the scientific method behind evolution and repelled by its inevitable outworking in nihilistic evolutionism. For Lewis, no matter what might have been scientifically true about natural selection, there was nothing but fantastical conjecture behind Evolution's materialist "creation story" purporting to provide an alternative to the biblical Creation story. To

Lewis's mind, this "Myth" of Evolution was nothing more than the story-telling of imaginative evolutionists, in which...

> [the cosmos was preceded by] the infinite void and matter endlessly, aimlessly moving to bring forth it knows not what. Then by some millionth, millionth chance—what tragic irony!—the conditions at one point of space and time bubble up into that tiny fermentation which we call organic life. [That organic life] spreads, it breeds, it complicates itself... from the amoeba up to the reptile, up to the mammal. [Finally,] there comes forth a little, naked, shivering, cowering biped, shuffling, not yet fully erect, promising nothing: the product of another millionth, millionth chance. His name in this Myth is Man.
>
> [Eventually] he has become true Man. He learns to master Nature. Science arises and dissipates the superstitions of his infancy. More and more he becomes the controller of his own fate. [Then after man becomes a race of demi-gods controlling all things, Nature returns with a vengeance and life is] banished without hope of return from every cubic inch of infinite space. All ends in nothingness.[346]

Lewis had no doubt that the Myth of evolutionism was absurdly out of touch with reality, including especially the spiritual reality which gives meaning to life. For years, Lewis contented himself with the thought that the problem was not with the "science" of Evolution but only with its overly-imaginative popularizers. Over time, Lewis began to see a closer connection between the Method and the Myth, and even penned a wry "hymn" lampooning Darwinism's inherent purposelessness.[347]

Lead us, Evolution lead us,
Up the future's endless stair…
Groping, guessing, yet progressing,
Lead us nobody knows where.

Lewis was not alone in his disquiet about a scientific theory that seemed so intrinsically devoid of purpose. But when staunch supporters of Evolution (such as George Bernard Shaw) attempted to imbue a purely materialist process with meaning and purpose (in what became known as "emergent evolution"), Lewis was having none of it.[348] Using sharp language, Lewis castigated the futility of attempting a "third way" to meld mindless materialism with any pretense of some designing Life-Force, which Lewis argued could be none other than "a God."[349] In *Mere Christianity*, Lewis asks: "What is the sense in saying that something without a mind 'strives' or has 'purposes'?"[350] Not exactly a rebuke to theistic evolution, but closely approaching it. *What is the sense in saying that God chose to use a process incapable of design as his means of achieving design?*

So, was Lewis a Darwinist? By no means steadfastly. Was he a creationist? Yes, if by that you mean he recognized God as the Creator, but not to the point of excluding a role for natural selection in the formation of species, perhaps even including humankind. Was Lewis, therefore, a theistic evolutionist or evolutionary creationist? Not by any self-description, and likely not if directly asked whether God might have used a mindless process to bring about his purposed (and empirically-obvious) design. Such an unseemly half-way house defied all Lewisian logic.

In truth, Lewis struggled with settling on any fixed position. By his own confession, his belief in Evolution was "of the vaguest and most intermittent kind."[351] Understanding why that should be the case with so prodigious a thinker and writer as Lewis may help explain why, even today, so many believers find themselves in such a muddle over Evolution. In the words of John West,

> Lewis eventually came to better understand just how intertwined evolution as a scientific theory was with what he had called evolutionism. Much of Lewis's growing awareness was likely due to his 16-year correspondence with Bernard Acworth, a leader in Britain's Evolution Protest Movement. Starting in the mid-1940s, Acworth began sending Lewis books and essays critical of Darwin's theory, materials which Lewis read and retained for his private library.
>
> [In 1951,] Acworth sent him a lengthy manuscript critical of evolution, and Lewis wrote back that he had 'read nearly the whole' of it. Acworth's manuscript hit home. 'I must confess it has shaken me,' Lewis wrote.[352]

In what way had Acworth's manuscript shaken the great Christian apologist? Not so much about evolution theory itself, but "*in my belief that the question was wholly unimportant.*" By any measure, Lewis's admission is huge; and so very telling about believers today who seem not to have grasped the full significance of their commitment to scientific evolution. Have they failed to fully appreciate what's at stake?

Nothing, absolutely nothing, could be more important than the question of origins, which Lewis finally came to realize (with what appears to be shock, if not regret, that it had taken so long). *It is on the question of origins that all of life's issues hang. Get origins wrong, and life becomes a meaningless Myth.*

Of course, evolutionary creationists would wholeheartedly agree, insisting with all their might that God created life and instilled within it both meaning and spiritual significance. Which explains why, for them (aside from their unflagging acceptance of what they believe to be unassailable evolution science), God's method of creation is wholly unimportant.

Unfortunately, despite all the theological clothes you might dress it in, the *method* of evolution is not only fatally flawed, but (not unlike the "built-in bias" proposed by theistic evolutionists) has its own "built-in bias"—a materialist bias inexorably producing a materialist *message* deleterious to faith. It's a God-denying message that C.S. Lewis was loathe to accept. A message it seems Lewis finally twigged was baked into evolution theory from the very start, no matter how much miraculous "God sauce" you pour over it.[353]

Hence, Lewis's response to Acworth that "…you may be right in regarding [Evolution] as the central and radical lie in the whole web of falsehood that now governs our lives…."[354] Say again? *Evolution* (not just evolutionism), *the central and radical lie.* For Lewis, finally, at last, dawns the light!

C.S. Lewis's evolution Odyssey ought to be a cautionary tale. To accommodate the theory is to accommodate its materialist, meaningless DNA. Faith cannot play host to such a destructive parasite and survive intact. If ever one demanded proof of that proposition, it can easily be seen in what happened to faith in Britain when the Church embraced Darwin and his Grand Theory….

When the Church Buried Darwin…
A Case Study in Compromise

Charles Robert Darwin died on April 19, 1882, at his home in Kent. His funeral was held a week later on April 26, and Darwin's body was interred in a place of honor at Westminster Abbey. The funeral was attended by fellow scientists, philosophers, naturalists, and dignitaries of all sorts–including, of course, leading members of the Anglican Church.

It did not seem to matter to the Church that Darwin's famed theory was problematic for the existence (or nature) of God; was troubling in its implications for morality, an immortal soul, divine judgment, and the afterlife. Or that it was laden with potential for undermining faith in Scripture, the Church, and Jesus Christ himself.

Nor did it seem to matter that Darwin had not attended church from the day he buried his beloved daughter, Annie; nor that by the end of his life Darwin had become an agnostic with no pretense of faith in a personal God; nor that Darwin would have made no claim to Jesus Christ being his Lord and Savior. The Church was more than happy to say grace over the man who was Britain's greatest naturalist—with few, if any, questions asked.

As indelicate as it may seem to say so, when the Church of England buried Darwin at Westminster Abbey, it quietly and imperceptibly dug its own grave. Not that the Church wasn't already spiritually moribund in many respects, but there is a very real sense in which the Church's celebration of Darwin's life and work was a significant nail in the coffin of its spiritual influence in Britain.

Not surprisingly, Darwin's bold new idea was initially met with official outcry and dismay by the Church. The debates were infamously fierce. One in particular, lasting four hours at Oxford University's Museum Library, pitted Bishop Samuel Wilberforce against Thomas Huxley and Joseph Hooker. But in time, many Anglican theologians in the nineteenth century positively embraced evolution. It was a time of discovery, exploration, and scientific enthusiasm. It was a time of progressive, enlightened thinking.

Consistent with Evolution's ever-widening acceptance over time, it is now so fully endorsed by the Church of England that, in 2009—celebrating two hundred years from Darwin's birth—the Church issued England's favorite son a formal apology, as told in this news clipping:

> The Church of England is to apologise to Charles Darwin for its initial rejection of his theories, nearly 150 years after he published his most famous work. The Church of England will concede in a statement that it was over-defensive and over-emotional in dismissing Darwin's ideas. It will call "anti-evolutionary fervour" an "indictment on the Church".

> The apology, which has been written by the Rev Dr Malcolm Brown, the Church's director of mission and public affairs, says that Christians, in their response to Darwin's theory of natural selection, repeated the mistakes they made in doubting Galileo's astronomy in the 17th century.
>
> The statement will read: "Charles Darwin: 200 years from your birth, the Church of England owes you an apology for misunderstanding you and, by getting our first reaction wrong, encouraging others to misunderstand you still. We try to practise the old virtues of 'faith seeking understanding' and hope that makes some amends."[355]

That's some evolution in the Church's thinking! But still no apparent concern about the negative spiritual implications of Darwin's theory. Or about Darwin's own personal lack of faith. And how quaintly odd, addressing the apology directly to Darwin himself as if, somewhere in the Great Beyond, Charles Darwin could hear, see, or somehow sense the apology! If there *is* a Great Beyond, the home of the soul, why would the Church apologize for wrongly reacting to a theory that fundamentally excludes both the soul and its other-worldly home? And what in Christian doctrine would suggest eternal approbation for someone who knowingly turns his back on God...and provides the perfect vehicle for countless others to travel down the same road to unbelief?

Regarding the gratuitous Galileo reference (as if there were a parallel!), is the Church so blind as to miss the *real* parallels? From antiquity, Aristotle had refuted heliocentricity (a sun-centered solar system), and by Galileo's time, virtually every major thinker subscribed to the geocentric (Earth-centered) view popularized by Ptolemy, the Greek mathematician and astronomer. So, Galileo upset, not just the church, but what then passed for *the entire scientific community*—both the pious and the impious. Ptolemy's hypothesis was the entrenched Darwinism of his day—for centuries! Most importantly, it hadn't derived from

Scripture, but from pagan (human) reasoning, which the Church foolishly baptized with legitimacy. Sound familiar?

Here's something else to ponder. If the Church could get a major scientific hypothesis so terribly wrong *once*, why should we think it couldn't get another scientific hypothesis terribly wrong *again*? And the Galileo debate didn't even have the slightest implications regarding human nature, immortality, transcendent morality, Christian doctrine, or social issues of any kind.

Is it any wonder how the Church of England has lost its moral authority in "Christian Britain?" Sadly, it's called *secularist compromise* and *spiritual abdication*. The same secularist compromise and spiritual abdication so intrinsic to theistic evolution. Take it from the experience of the Church of England: commingling religion and Darwinism is a death knell for faith.

But back to basics. If Darwin's fatally-flawed Grand Theory is bad science, why would any person of faith wish to affirm its scientific validity? Or dare attempt to insinuate it into God's creative mind? Or lend the slightest credence to its built-in, faithless message?

The words of the ancient prophet come to mind: "How long will you waver between two opinions?"

Recap

- If the "cake-and-eat-it-too" approach of theistic evolutionists is so obvious you can't miss it, occasionally subtler versions appear (as when you hear words like "built-in bias," or "prefigured," or "pre-ordained," or even references to "natural laws," which necessarily require a purposing Lawmaker), putting us on alert that the writer is *thinking* "evolutionary creation" even if he doesn't *say* it outright.

- Theistic evolution sometimes cloaks itself in robes of theological scholarship, hoping to be a neutral arbiter between creation and evolution by declaring that the "creation story" in Genesis was never intended to be anything more than a theological construct, answering the question of *why* God created, not *how* he created.

- Religious theists who so eagerly line up with secular evolutionists are unwittingly lending credence, not simply to a fatally-flawed scientific theory, but to the "secular DNA" inherent within evolution theory.

- The evolution in C.S. Lewis's fluctuating views on evolution theory gives insight as to why even the brightest of the bright would entertain the notion of God-ordained evolution. Yet it also highlights that the more one understands about the mechanisms of evolution, the more it's not just the romanticized Evolution Story that is so fantastical, but the flawed science behind it.

- If, as this book proposes, Darwin's secret sex problem is fatal to the microbe-to-man Evolution Story, merely adding the word "theistic" neither saves that fictional story nor enhances it, but only serves to detract from the Biblical Story and its profound implications for the Human Story.

Question: Now that you've all but finished the book, can you see that Darwin's "secret sex problem" is fatal to both classic, textbook evolution and to any futile attempt to meld it with divine Creation?

EPILOGUE

In questions of science, the authority of a thousand
is not worth the humble reasoning of a single individual.
—Galileo Galilei

So, what do you think? What will you do with this information? If you teach biology in primary or secondary education, how will you teach it in the future? If you are a college professor, will your lectures be different? If you are an evolutionist, might you need to reconsider which aspects of evolution you can continue to accept and which aspects you should jettison? If you consider yourself a theistic evolutionist or evolutionary creationist, does this book give you pause to reconsider any attempt to conflate biblical Creation and microbe-to-man evolution?

One of the interesting things about major shifts in scientific paradigms is that there's an observable pattern of 1) orthodoxy, 2) challenge to orthodoxy, 3) rejection of the challenge, and 4) finally—at some tipping point—acceptance of the challenge, which itself becomes the new orthodoxy. It happened in the Copernican Revolution, of course, and certainly in the Darwinian Revolution. It's not as if no one before Copernicus had ever questioned the universally-accepted geocentric assumption. The problem was that there seemed to be no viable alternative to the flawed geocentric hypothesis. Only when Copernicus offered a robust, scientifically-provable alternative did the previous hypothesis collapse faster than a planet hurtling around the sun.

For a variety of reasons, the Darwinian Revolution has encountered far more resistance and prolonged objection. For starters, the revolutionary heliocentric paradigm may have overturned an accepted tenet of religious thought in its day, but Darwinism went straight for the jugular. Simultaneously, it dethroned not simply the nobility and superiority of humankind but the creative power and purpose of the Almighty. If for no other reason, to this day Darwin's Grand Theory is still under attack.

Despite this ongoing opposition, microbe-to-man evolution is unquestionably the most widely-accepted working theory of science. In short, the Evolution Story popularized by Darwin is today's entrenched orthodoxy. Suppose then that the thesis of this book is right—that Darwin's Grand Theory is fundamentally flawed for being unable to explain the appearance of sex. Now take another giant leap and suppose that many among the scientific community actually begin taking the problem seriously and start rethinking the Darwinian hypothesis. Is there even a remote chance that the day could come when there is yet another scientific revolution? A day when orthodox science rejects the origin of all species from a common ancestor by a process of gradual evolution?

If history has a message for us, it is that—absent an acceptable option—folks will stick with even a bad theory to the dying end. Something like, "better the devil we know," only this one has a wry twist: Better a flawed scientific explanation than an explanation we have other reasons (primarily philosophical) to reject. Simply to argue that there are no viable *scientific* alternatives doesn't mean there are no (altogether reasonable) *non-scientific* alternatives. Love is hardly capable of scientific proof, but who would dare deny its existence?

One thing is clear: Fallacies in a theory never depend on either the availability or lack of viable alternatives. If a fallacy is genuine, it is still a fallacy even if there is no known alternative. Keeping it simple, if a vehicle is broken down and won't run, it is broken down whether or not another vehicle is available.

The only question, then, is whether Darwin's Grand Theory can hold together in the face of evolution's catastrophic sex problem. If it can't, then it should be rejected by the scientific community even if there were absolutely no alternative explanation waiting in the wings. If the problem of sex is *in scientific fact* fatal to Darwin's Grand Theory, are evolutionists really content to say they are sticking with a fatally-flawed theory simply for lack of better options?

So, is there a better option? At the very least, if evolution's irredeemable sex problem nullifies Darwin's Grand Theory, the reasonable mind would not be amiss to consider some external agency as the cause of this wondrous world in which sex is both prolific and profound. What else (or who else) would that external agency be? Certainly not the compromised God of "evolutionary creation" so beholden to Darwin.

The most obvious alternative—instantaneous Creation— is, of course, the very alternative the scientific community is so adamant to reject, and the one about which there is increasing embarrassment among scientifically-entranced believers. The question is, do we have sufficient reason to reject an altogether plausible explanation simply because we don't want to accept its wider implications? How we answer that question probably says less about what explanation of life we ultimately choose to believe than what that choice says of us and our deepest bedrock values.

There is one group I didn't mention in the opening paragraph—today's young people. As tomorrow's thinkers, teachers, and leaders, they are the most important group of all. So, if you are a student, you need to ask your teacher or professor about evolution's sex problem. Ask them out of class (where they're more prone to open up) or even in class where appropriate. Just don't let them get away with dodgy (usually circular) arguments; or overly-hyped "evidence" (about the gaping fossil record, and anatomical similarities, and computerized mathematical studies), or supercilious rhetorical flourishes (where ginormous gaps are leaped with the flair of Superman's cape).

If you are a believer, you also need to ask your church leaders if they've ever thought about evolution's sex problem (and maybe whether they've bought off on the notion that God chose to create this marvelous universe using a flawed scientific process). At home, switch roles with your parents to have "the talk"...about the birds and the bees. Odds are, they don't yet know what you know!

Most of all, it's time you had a frank talk with your fellow students and friends about sex and evolution. You don't need to know all the technical ins and outs of mitosis and meiosis. Just mention "half a penis, half a vagina," and watch light bulbs come on as billions of mythical evolutionary years go flying out the window! It's simple. They'll get it.

It's becoming increasingly clear that this generation no longer trusts politicians or other authority figures. Just listen to all the raucous protests! What, then, is the most trusted social institution today? None other than the scientific community. But when the scientific community is feeding us the biggest scientific lie ever told, it's time we turned our anger on something truly worth protesting.

With the writing of *The Origin of Species*, Darwin dramatically changed his world...and ours. It's time to change it back. What we're needing here is nothing short of a revolution. *An evolution revolution!* A revolution that not only turns the fictional Evolution Story on its head, but revolutionizes science itself.

Indeed, what we need is a revolution that spawns a whole new generation of truth-seeking and truth-telling—about where we've come from, about how we really got to be humans, and, most important, about why on earth we are here.

GLOSSARY

Allele, a variation of a gene providing for a particular genetic trait.

Asexual, where replication of an organism occurs by mitosis or other means not involving the production of sex cells or the process of fertilization.

Autosomal chromosome, any chromosome not related to sex determination.

Binary fission, how prokaryotic cells divide into two daughter cells.

Bounded evolution, development of genetic changes below the speciation threshold.

Budding, outgrowth from an organism, which separates without sexual reproduction to form a new "individual" (as can be found in yeast). Budding occurs commonly in some invertebrate animals such as corals and hydras. In hydras, a bud forms that develops into an adult, which breaks away from the main body; whereas in coral budding, the bud does not detach and multiplies as part of a new colony.

Chromatid, the two thread-like strands into which a chromosome divides during cell division.

Chromatin, uncoiled genetic material composed of DNA and proteins that condense to form chromosomes.

Chromosome, a structure in the nucleus of living cells consisting of nucleic acid and protein which carry genetic information.

Cloning, asexual replication from an ancestor to which the offspring is identical.

Coevolution, the supposed (but unsupported) mutual evolution between two species, as in predator and prey, or between male and female of a species. Sometimes referred to as "tandem evolution."

Common descent (common ancestor, common origin, common progenitor), the linchpin of evolutionary theory whereby all higher species have evolved from one or more lower-level prototypes, common to all species arising therefrom.

Conjugation, the transfer of DNA between two cells, as in the temporary union of two bacteria or unicellular organisms in which genetic material is swapped.

Crossing over, the point in meiosis (sexual reproduction) when portions from two chromosomes wrap around each other, resulting in a genetic exchange between the chromosomes (contributing to genetic variation among offspring).

Cytokinesis, the separation of the cytoplasm of a parent cell to form two daughter cells (both in mitosis and meiosis) after telophase.

Deletion, occurs when a piece of a chromosome is lost.

Dioecy, where male and female reproductive organs are divided from conjoined sexes, or exist in separate individuals.

Diploid, condition in which two sets of homologous chromosomes are present.

DNA, deoxyribonucleic acid, the self-replicating material which carries the genetic information.

Double helix, the two twisting strands of DNA, held together by hydrogen bonds.

Enzyme, a protein acting as a catalyst to promote a particular biochemical reaction.

Estrus, the time when, in most animals, the female is more receptive to the male's sexual overtures.

Eukaryotic cell, a complex cell in which much of the genetic material is contained within a nucleus

Evolution, at its simplest, modifications in form and function of living organisms. Typically confused with the Grand Theory of Evolution (either Darwinian or neo-Darwinian), according to which simple, lower life forms are said to have gradually changed over millions of years through a process of "natural selection" into complex, higher life forms, from microbes to man.

Evolutionary creation (See theistic evolution)

Fission, process in which a cell or unicellular organism divides into two cells or unicellular organisms.

Fragmentation, the breaking of the body into two parts with subsequent regeneration. If the organism is capable of fragmentation, and the part is big enough, a separate individual will regrow.

Fossil record, remains or impressions of organisms found preserved, typically in rocks.

Gamete, a mature male or female germ cell capable of uniting with a germ cell from the opposite sex during fertilization.

Gene, a coded unit of heredity information which, when expressed genetically, influences metabolism and often influences a visible trait of the offspring.

Genetic drift, a type of evolution said to occur by the chance disappearance of genes, typically in small groups of organisms.

Genus, a category within the taxonomic hierarchy to which one or more *species* belongs.

Geocentric, pertaining to the model articulated by Aristotle and, especially, Ptolemy that places the earth in the center of the universe with the sun revolving around the earth.

Haploid, a genetic condition in which only one member of each pair of homologous chromosomes is present.

Heliocentric, pertaining to the model of the earth and the planets rotating around the sun, as championed by Copernicus, Kepler, and Galileo, overturning the geocentric model of earlier philosophers.

Hermaphrodite, an organism having both male and female organs.

Hominins (hominids), two categories into which humans have been classed with either presumed precursors or other evolutionary-related species, such as great apes (chimpanzees, gorillas, and orangutans) and any extinct species descended from their supposed common ancestor.

Homologous chromosomes, chromosomes that usually resemble one another in size and that carry genetic information for the same traits.

Homology, anatomical similarity, used as an argument for evolution, but subject to interpretation.

Intelligent design, the belief that the universe and humankind are the products of intelligent design rather than the blind chance normally associated with natural selection and evolution.

Invertebrates, animals without backbones, representing 95% of all species.

Irreducible complexity, the central argument of the Intelligent Design position that evolution cannot account for those biological and chemical systems within organisms which require a package of interacting parts in order to function. If because of evolution's gradualism a single part is missing, any system dependent on that part could not possibly function.

Maintenance of sex, pertaining to explanations as to why sexual reproduction is advantageous to maintaining the genetic line of sexually-reproducing organisms (in contrast to the *origin* of sex).

Meiosis (meiotic), the process necessary for sexual reproduction by which an organism produces gametes by reduction of chromosome number to the haploid number and genetic recombination of chromosomes (in contrast to the exact-copy process of mitosis).

Mitosis (mitotic), a process involving DNA and chromosome replication and nuclear division which, when followed by cytokinesis (cytoplasmic division), allows cell proliferation, growth, and in some species, asexual replication. (Usually understood in contrast to the process of sexual reproduction, or meiosis.)

Mutation, a typically random event that causes a change in the coded sequence that represents the information conveyed by a gene.

Natural selection, the process whereby organisms either survive or are eliminated depending on their relative "fitness" to survive within their environment.

Neo-Darwinism, also known as the "Modern synthesis," is an amalgam of modern genetics, taxonomy, systematics, paleontology, botany, zoology, and ecology which has built on Darwin's original theory.

Nucleic acid, a polymeric molecule consisting of a long chain of subunits called nucleotides serving as the genetic material for living organisms. Nucleic acids are of two major types, DNA (deoxyribonucleic acid) and RNA (ribonucleic acid).

Oogenesis, process by which eggs are formed in the female sexual structures.

Organelles, distinct membrane-bound structures such as mitochondria and chloroplasts within a living cell.

Oviduct. the fallopian tube connecting the ovary and uterus.

Parthenogenesis, reproduction from the female ovum without male fertilization.

Pheromones, chemical signals given off as scents, often associated with the process of mating.

Phyla, a category of taxonomy that is above "class" and below "kingdom."

Ploidy, the number of sets of chromosomes in a cell, or in an organism's cells.

Polar body, a cell produced then discarded during the process of meiosis and gamete maturation.

Primary oocytes, cells created in the first phase of meiosis, but which sit idle until the female enters puberty.

Prokaryotic cell, a cell in which the nuclear material is not surrounded by a nuclear envelope (in contrast to eukaryotic cells which have nuclei).

Protein, a polymeric molecule composed of amino acids and functioning in cellular and organismal structure and as enzymes in metabolism.

Protists, mostly single-celled organisms, like protozoa, algae, and fungi.

Punctuated equilibrium, a model theorizing that evolution occurs in quick bursts, also known as saltational leaps.

RNA, ribonucleic acid, a type of nucleic acid which determines amino acid sequence during protein synthesis when the DNA code of genes is expressed.

Recombination, the process by which the genes (alleles) at different locations in two parental individuals become shuffled in the offspring.

Saltational evolution, the belief that evolution occurs in jumps and starts rather than gradually. See also "punctuated equilibrium."

Secondary oocyte, an oocyte in which the first division of meiosis (meiosis I) is completed.

Seminal vesicles, sources of fluids entering the ejaculatory duct along with sperm.

Seminiferous tubules, site where sperm (spermatozoa) are produced.

Sexual dimorphism, recognizable difference in form between males and females of any given species, as in size or color.

Sexual reproduction, a process in which offspring are produced by a process of meiosis which occurs in separate sexes and is followed by fertilization.

Sexual selection, process related to evolution which is said to lead to certain characteristics (typically male) being naturally selected as beneficial to acquiring mates (as in deer antlers and peacock tails).

Species, a group of organisms capable of interbreeding and exchanging genes.

Spermatids, immature sperm (spermatozoa) in the waiting phase, before launching into the race for the ovum.

Spermatozoa (sperm), the male gamete which unites with the female egg in meiosis.

Taxonomy, the naming and classification of organisms according to shared characteristics.

Theistic evolution (evolutionary creation), the belief that God created humankind progressively over millions of years through a process of evolution guided and sustained by divine providence, purpose, and design.

Transduction, occurs when genes from one cell move to another cell.

Unbounded evolution, evolution from one species to another, resulting in microbe-to-man progression.

Urethra, the duct which ejects both urine and semen in males, and urine in females.

Vas deferens, the duct connecting the epididymis to the urethra.

Vertebrates, animals having backbones.

Vestigial characters, organs or features of organisms thought to have degenerated during evolution from a prior functional organ or feature, such as the human appendix and the os coccyx.

INDEX

D

E

F

G

H

I

J

K

L

M

N

O

P

Q

R

S

NOTES

1. Richard Dawkins, *The Selfish Gene* (Oxford, New York: Oxford University Press, 30th Anniversary Edition, 2006, Reprinted 2009), 43
2. Dawkins, *Selfish Gene*, 44
3. Dawkins, *Selfish Gene*, 44
4. Richard Dawkins, *The Blind Watchmaker* (New York, London: W.W. Norton & Company, 1986), 268
5. Richard E. Michod, *Eros and Evolution—A Natural Philosophy of Sex* (Reading, MA: Addison-Wesley Publishing Company, 1995), 172-173
6. Mark Ridley, *The Problems of Evolution* (Oxford, New York: Oxford University Press, 1985), 90
7. Ridley, *Problems*, 124
8. Carl Zimmer, *Evolution—The Triumph of an Idea* (New York: Harper Perennial, 2006), 187
9. "I Hear You Knocking," first published in 1955 and recorded by Smiley Lewis, was released in 1970 by Dave Edmunds.
10. Alan F. Dixson, *Sexual Selection and the origins of Human Mating Systems* (Oxford University Press, 2009), 73
11. Dixson, *Mating Systems*, 73
12. Dixson, *Mating Systems,* 68
13. Dixson, *Mating Systems,* 65
14. Charles Darwin, *The Origin of Species* (New York: Signet Classics, 150th Anniversary Edition, 2003), 150
15. Zimmer, *Triumph*, 229
16. Zimmer, *Triumph*, 227-232
17. Dixson, *Mating Systems,* 50
18. Dixson, *Mating Systems,* 50
19. See Ridley, *Problems*, 117
20. Ridley, *Problems*, 116
21. Francis S. Collins, *The Language of God: A Scientist Presents Evidence for Belief* (New York: Free Press, 2006), 138
22. John Maddox, *What Remains to be Discovered* (New York: Free Press, 1999), 254-255

23 Lynn Margulis and Dorion Sagan, *Origins of Sex* (New Haven: Yale University Press, 2nd Printing, 1986), 37
24 Margulis and Sagan, *Origins of Sex*, 107
25 Margulis and Sagan, *Origins of Sex*, 146
26 Margulis and Sagan, *Origins of Sex*, 146
27 Michael Pitman, *Adam and Evolution* (London: Rider, 1984), 67
28 Arthur Koestler, *The Ghost in the Machine* (London, New York: Picador, 1975), 129
29 Ridley, *Problems*, 34
30 See Lane P. Lester and Raymond G. Bohlin, *The Natural Limits to Biological Change* (Grand Rapids: Zondervan Publishing, 1984), 58, 68, 170
31 Pitman, *Adam and Evolution*, 70
32 Zimmer, *Triumph*, 97
33 Rene Fester Kratz and Donna Rae Siegfried, *Biology For Dummies* (Hoboken: John Wiley & Sons, 2010), 304
34 Mark Jerome Walters, *The Dance of Life—Courtship in the Animal Kingdom* (New York: Arbor House, William Morrow and Company, 1988), 71
35 Pitman, *Adam and Evolution*, 112
36 Peter Brent, *Charles Darwin* (Feltham, England: Hamlyn Publishing Group, 1981), 284
37 Stephen Jay Gould, *Wonderful Life* (New York: W.W. Norton & Company, 1989), 60, 64
38 Brent, *Charles Darwin,* 212
39 Denton, Michael, *Evolution, A Theory in Crisis* (Chevy Chase, MD: Adler & Adler Publishers, 1986), 81-82
40 Carl Zimmer, *Triumph*, 56
41 T. Dobzhansky, F.J. Ayala, G.L. Stebbins, and J.W. Valentine, *Evolution* (San Francisco: W.H. Freeman and Co, 1977), 271-76 [Denton, *Theory in Crisis*, 82]
42 Gerald Schroeder, *The Science of God* (New York: Free Press, 2009), 32
43 Denton, *Theory in Crisis*, 86
44 Ridley, *Problems*, 7
45 Ridley, *Problems*, 56-58
46 Darwin, *Origin of Species*, 39
47 Darwin, *Origin of Species*, 94
48 Darwin, *Origin of Species*, 159
49 Brent, *Charles Darwin,* 284
50 Brent, *Charles Darwin,* 287
51 Darwin, *Origin of Species*, 433
52 Zimmer, *Triumph*, 140
53 Zimmer, *Triumph*, 141
54 Zimmer, *Triumph*, 145-147

55 Carl Zimmer and Douglas J. Emlen, *Evolution—Making Sense* (Greenwood, CO: Roberts and Company, 2013), 402
56 Zimmer and Emlen, *Evolution—Making Sense,* 400
57 Ridley, *Problems*, 4-5
58 Charles Darwin, *Descent of Man* (New York: Barnes and Noble, Inc., 2004), 140
59 Marshall, Perry, *Evolution 2.0—Breaking the Deadlock Between Darwin and Design* (Dallas: Benbella Books, 2015), 298
60 Marshall, *Evolution 2.0*, 42-44
61 Marshall, *Evolution 2.0*, 297
62 Marshall, *Evolution 2.0*, 59
63 Marshall, *Evolution 2*.0, 217
64 Marshall, *Evolution 2*.0, 137
65 Marshall, *Evolution 2*.0, 143
66 Romans 1:25
67 Marshall, *Evolution 2*.0, 287
68 Jerry A. Coyne, *Why Evolution is True* (Oxford: Oxford University Press, 2009), 185
69 Coyne, *Why Evolution*, 6
70 Coyne, *Why Evolution*, 187
71 Coyne, *Why Evolution*, 190
72 Coyne, *Why Evolution*, 192
73 Paul A. Nelson "Five Questions Everyone Should Ask about Common Descent," in *Theistic Evolution*, ed. J.P. Moreland, Stephen C. Meyer, Christopher Shaw, Ann K. Gauger, Wayne Grudem (Wheaton: Crossway, 2017), 416
74 Michael Behe, *Darwin's Black Box* (New York: Touchstone, Simon & Schuster, 1996), 39
75 Dixson, *Mating Systems,* 92
76 Dixson, *Mating Systems,* 59
77 Dixson, *Mating Systems,* 79
78 Denton, *Theory in Crisis*, 209
79 Stephen Jay Gould, "The Problem of Perfection, or How Can a Clam Mount a Fish on Its Rear End?" in *Ever Since Darwin* (New York: W.W. Norton, 1979, reissued 2007), 104
80 Denton, *Theory in Crisis*, 202
81 Walters, *Dance of Life*, 14
82 Denton, *Theory in Crisis*, 218
83 Zimmer, *Triumph*, 187
84 Darwin, *Origin of Species*, xiv
85 Denton, *Theory in Crisis*, 219
86 R. J. Tillyard, *The Biology of the Dragonfly* (Cambridge: Cambridge University Press, 1917), 215

87 Denton, *Theory in Crisis*, 227
88 Regina Nuzzo, "Female insect uses spiky penis to take charge," Nature, April 17, 2014 [http://www.nature.com/news/female-insect-uses-spiky-penis-to-take-charge-1.15064]
89 Walters, *Dance of Life*, 43
90 Colin Barras, "The Twisted World of Sexual Organs—How evolution solved the problem of conception," 2014, http://www.bbc.co.uk/bbc.com/earth/bespoke/story/20140908-twisted-world-of-sexual-organs/index.html
91 Dixson, *Mating Systems,* 99
92 Dixson, *Mating Systems,* 106
93 Gould, *Wonderful Life*, 60
94 Olivia Judson, *Dr. Tatiana's Sex Advice to All Creation* (New York: Henry Holt and Company, 2002), 138
95 Judson, *Dr. Tatiana's Sex Advice*, 234
96 Judson, *Dr. Tatian's Sex* Advice, 12
97 Dixson, *Mating Systems,* 53
98 Dixson, *Mating Systems,* 58
99 Dixson, *Mating Systems,* 52
100 Dixson, *Mating Systems,* 59
101 Dixson, *Mating Systems,* 71
102 Dixson, *Mating Systems,* 78
103 George C. Williams, *Sex and Evolution* (Princeton: Princeton University Press, 1975), 124
104 Darwin, *Origin of Species*, 502
105 Zimmer and Emlen, *Evolution—Making Sense*, 320
106 Richard Dawkins, *The Extended Phenotype* (Oxford: Oxford University Press, 1982), 106
107 Adam S. Wilkins and Robin Holliday, "The Evolution of Meiosis From Mitosis," *Perspectives*, Anecdotal, Historical and Critical Commentaries on Genetics, ed. By James F. Crow and William F. Dove, *Genetics* 181:3-12 (January 2009)
108 Hamilton, W.J., *Narrow Roads to Gene Land: Evolution of Sex*, Vol. 2. (Oxford: Oxford University Press, 1999), 419
109 Mark Ridley, *The Cooperative Gene* (New York: Free Press, 2001), 226-227
110 Adam S. Wilkins and Robin Holliday, "The Evolution of Meiosis From Mitosis," *Perspectives*, Anecdotal, Historical and Critical Commentaries on Genetics, ed. By James F. Crow and William F. Dove, *Genetics* 181:3-12 (January 2009)
111 Aanen D, Beekman M, Kokko H. 2016 Weird Sex: the unappreciated diversity of sexual reproduction. *Phil. Trans. R. Soc. B* **371**: 20160262. http://dx.doi.org/10.1098/rstb.2016.0262
112 Daniel J. G. Lahr, Laura Wegener Parfrey, Edward A. D. Mitchell, Laura A. Katz, and Enrique Lara, "The chastity of amoebae: re-evaluating evidence for

sex in amoeboid organisms," Proceedings of the Royal Society B, published online before print 23 March 2011 doi: 10.1098/rspb.2011.0289 Proc. R. Soc. B 22 July 2011 vol. 278 no. 1715 2081-2090

113 Richard E. Michod and Bruce R. Levin, eds., *The Evolution of Sex* (Sunderland, MA: Sinauer Associates, 1988), citing Bernstein, Hopf and Michod, 1411.

114 Graham Bell, *The Masterpiece of Nature: The Evolution of Genetics and Sexuality* (Berkeley, CA: University of California Press, 1982), 34

115 Williams, *Sex and Evolution*, 4

116 Williams, *Sex and Evolution*, 4

117 Zimmer and Emlen, *Evolution—Making Sense,* 422

118 Walters, *Dance of Life*, 6

119 Ridley, *Cooperative Gene*, 129

120 Nick Lane, *Life Ascending, The Ten Great Inventions of Evolution* (New York: W.W. Norton & Company, 2009), 140

121 Margulis and Sagan, *Origins of Sex*, xix

122 Margulis and Sagan, *Origins of Sex*, 2

123 Margulis and Sagan, *Origins of Sex*, 3

124 Margulis and Sagan, *Origins of Sex*, 9

125 Margulis and Sagan, *Origins of Sex*, 3

126 Bell, *Masterpiece of Nature*, 19-26

127 See Zimmer and Emlen, *Evolution—Making Sense,* 331

128 Megan Scudellari, "The Sex Paradox," The Scientist Magazine, July 1, 2014. [http://www.the- scientist.com/?articles.view/articleNo/40333/title/The-Sex-Paradox/]

129 Zimmer and Emlen, *Evolution—Making Sense,* 335

130 Megan Scudellari, "The Sex Paradox," The Scientist Magazine, July 1, 2014. [http://www.the- scientist.com/?articles.view/articleNo/40333/title/The-Sex-Paradox/]

131 Maddox, *What Remains*, 252-254

132 Maddox, *What Remains*, 254

133 Nicolas Money, *The Amoeba in the Room* (Oxford: Oxford University Press, 2014), xvii

134 Pitman, *Adam and Evolution*, 106

135 Darwin, *Origin of Species*, 174

136 Zimmer, *Triumph*, 59

137 Michod and Levin, *Evolution of* Sex, 46

138 Michod, *Eros and Evolution*, 63

139 Ridley, *Cooperative Gene*, 108

140 Paul Davies, *About Time—Einstein's Unfinished Revolution* (New York, London: Simon & Schuster Paperbacks, 1995), 250

141 Margulis and Sagan, *Origins of Sex*, 200-201

142 Martin Daly and Margo Wilson, *Sex, Evolution, and Behavior* (Belmont, CA: Wadsworth Publishing Company, 2nd Ed., 1983), 71
143 Michod, *Eros and Evolution*, 41
144 The TalkOrigins Archive, http://www.talkorigins.org/indexcc/CB/CB350.html
145 Bell, *Masterpiece of Nature,* 19
146 Ridley, *Cooperative Gene*, xi
147 Ridley, *Cooperative Gene*, 144
148 Ridley, *Cooperative Gene*, 161
149 Walters, *Dance of Life*, 6-7
150 Walters, *Dance of Life*, 11
151 Darwin, *Descent of Man*, 548
152 "Origin of Sex Pinned Down," livescience.com website, November 25, 2008
153 R.B. Spigler, K.S. Lewers, D.S. Main and T.L. Ashman, "Genetic mapping of sex determination in a wild strawberry, Fragaria virginiana, reveals earliest form of sex chromosome," Heredity Journal, December, 2008, 101, 507—517; doi:10.1038/hdy.2008.100; published online 17 September 2008
154 http://www.livescience.com/7613-origin-sex-pinned.html
155 Brent, *Charles Darwin*, 377-378
156 Darwin, *Origin of Species*, 42
157 Darwin, *Origin of Species*, 53
158 Margulis and Sagan, *Origins of Sex*, 3
159 Bell, *Masterpiece of Nature*, 91-101
160 Williams, *Sex and Evolution*, 15, 17, 37, 53
161 Leigh Van Valen, "A New Evolutionary Law," *Evolutionary Theory*, 1973, 1:1-30
162 Michod, *Eros and Evolution*, 38
163 Dawkins, *Selfish Gene*, 14
164 Dawkins, *Selfish Gene*, 19
165 Zimmer, *Triumph*, 166
166 Zimmer, *Triumph*, 167
167 Colin Patterson, "Are the Reports of Darwin's Death Exaggerated?", BBC Radio 4, 2 October, 1981 [See Pitman, *Adam and Evolution*, 45
168 Stephen Jay Gould, *The Panda's Thumb* (New York: W.W. Norton & Company, 1980), 189
169 See Brent, *Charles Darwin,* 298
170 Zimmer, *Triumph*, 7
171 See Zimmer, *Triumph*, 50
172 Brent, *Charles Darwin*, 298-299
173 Brent, *Charles Darwin*, 302-303
174 Brent, *Charles Darwin*, 299
175 Darwin, *Origin of Species*, 149

176 Brent, *Charles Darwin*, 288
177 Brent, *Charles Darwin*, 305
178 Darwin, *Origin of Species*, 99
179 Darwin, *Origin of Species*, 122
180 Kratz and Siegfried, *Biology For Dummies*, 81
181 Bell, *Masterpiece of Nature*, 84
182 Mark Anestis and Kellie Ploeger Cox, *5 Steps To A 5, AP Biology* (New York: McGraw-Hill, 2013), 147
183 Douglas J. Futuyma, *Evolution*, 3rd Ed. (Sunderland, MA: Sinauer Associates, Inc., 2013), 402
184 Williams, *Sex and Evolution* (entire book)
185 Dixson, *Mating Systems,* (entire book)
186 Ridley, *Cooperative Gene*, 108, 111
187 Ridley, *Cooperative Gene*, ix
188 Margulis and Sagan, *Origins of Sex*, 1
189 Margulis and Sagan, *Origins of Sex*, 205
190 Margulis and Sagan, *Origins of Sex*, 150
191 John Maynard Smith and Eörs Szathmáry, *The Major Transitions in Evolution* (New York: Oxford University Press, Reprinted 2008), 147-67
192 Bell, *Masterpiece of Nature*, 19-26
193 Bell, *Masterpiece of Nature*, 86
194 Bell, *Masterpiece of Nature*, 390
195 Joris Paul van Rossum, "On Sexual Reproduction as a New Critique of the Theory of Natural Selection," (Netherlands: Printforce, ISBN 978-90-9027200-9, NUR 738 Amsterdam 2012, with corrections 2014), 58
196 Darwin, *Origin of Species*, 62
197 Darwin, *Origin of Species*, 76-77
198 Zimmer and Emlen, *Evolution—Making Sense,* 499
199 Zimmer and Emlen, *Evolution—Making Sense,* 496
200 Darwin, *Origin of Species*, 97
201 Darwin, *Origin of Species*, 5
202 Lane, *Life Ascending*, 140
203 Lane, *Life Ascending*, 133
204 Lane, *Life Ascending*, 141
205 Lane, *Life Ascending*, 141
206 Lane, *Life Ascending*, 142
207 Lane, *Life Ascending*, 142
208 Lane, *Life Ascending*, 142
209 Lane, *Life Ascending*, 139
210 See Pitman, *Adam and Evolution*, 200
211 See Pitman, *Adam and Evolution*, 207

212 Van Rossum, "Sexual Reproduction," 143
213 Van Rossum, "Sexual Reproduction," 142
214 Darwin, *Origin of Species*, 176
215 Van Rossum, "Sexual Reproduction," 137-138
216 Van Rossum, "Sexual Reproduction," 136
217 Van Rossum, "Sexual Reproduction," 139
218 Zimmer, *Triumph,* xix
219 Gould, *Panda's Thumb*, 225
220 Charles Darwin, Letter to Joseph Hooker, 1844
221 Zimmer, *Triumph*, xxx
222 Zimmer, *Triumph*, 342-343
223 Zimmer, *Triumph*, 342-343
224 Pitman, *Adam and Evolution*, 74, citing Stephan Jay Gould, and Niles Eldredge, *Paleobiology*, vol. 3, Spring, 1977, 145
225 Walters, *Dance of Life*, 72
226 Judith Hooper, *Of Moths and Men: An Evolutionary Tale* (New York: W.W. Norton & Company, 2002), xix
227 Darwin, *Origin of Species*, 243
228 Darwin, *Origin of Species,* xx
229 Zimmer, *Triumph*, 156
230 Darwin, *Origin of Species*, 13-15
231 Darwin, *Origin of Species*, 78
232 Darwin, *Origin of Species*, xxviii
233 See Zimmer, *Triumph*, 43-44
234 Darwin, *Origin of Species*, 507
235 Darwin, *Descent of Man*, 48
236 Darwin, *Origin of Species*, 7
237 Darwin, *Origin of Species*, xxiv
238 Darwin, *Origin of Species*, 187
239 Darwin, *Origin of Species*, 175
240 Darwin, *Origin of Species*, 175
241 Darwin, *Origin of Species*, 176
242 Darwin, *Origin of Species*, 191
243 Brent, *Charles Darwin,* 270
244 Darwin, *Origin of Species*, 489
245 Darwin, *Origin of Species*, 506
246 Brent, *Charles Darwin,* 452
247 Darwin, *Origin of Species*, 160
248 BioLogos Foundation, www.biologos.org
249 Jerry A. Coyne, *Faith vs. Fact—Why Science and Religion Are Incompatible* (New York: Viking, Penguin Publishing Group, 2015), 16

250 William A. Dembski, *Intelligent Design—The Bridge Between Science and Theology* (Downers Grove: InterVarsity Press, 1999), 112
251 Dennis R. Venema and Scot McKnight, *Adam and the Genome* (Grand Rapids, Brazos Press, 2017), xii
252 Denis O. Lamoureux, *Evolutionary Creation—A Christian Approach to Evolution* (Eugene, OR: Wipf & Stock, 2008), 384
253 Lamoureux, *Evolutionary Creation*, 175
254 Lamoureux, *Evolutionary Creation*, 161
255 Lamoureux, *Evolutionary Creation*, 309
256 Dembski, *Intelligent Design*, 110
257 Lamoureux, *Evolutionary Creation*, 30-31
258 Collins, *Language of God*, 205
259 Collins, *Language of God*, 95-96
260 Coyne, *Faith vs. Fact*, 100-101
261 Brent, *Charles Darwin,* 452
262 Lamoureux, *Evolutionary Creation*, 62
263 Collins, *Language of God*, 200
264 Collins, *Language of God*, 74
265 Lamoureux, *Evolutionary Creation*, 30
266 Lamoureux, *Evolutionary Creation*, 31
267 Richard S. Beal, Jr., *The Grand Canyon, Evolution, and Intelligent Design* (Savage, MN: Lighthouse Christian Publishing, 2007), 18
268 Beal, *Grand Canyon*, 153-159
269 Beal, *Grand Canyon*, 163
270 Collins, *Language of God*, 200-201
271 Coyne, *Faith vs. Fact*, 140
272 Collins, *Language of God*, 188
273 Collins, *Language of God*, 201
274 Pope John Paul II, "Message to the Pontifical Academy of Sciences: On Evolution," Oct. 22, 1996
275 Lewis, C.S., *The Problem of Pain*, (New York: Simon & Schuster, 1996) 68-71
276 Lamoureux, *Evolutionary Creation*, 289
277 Lamoureux, *Evolutionary Creation*, 100
278 Lamoureux, *Evolutionary Creation*, 101
279 Lamoureux, *Evolutionary Creation*, 290-291
280 1 Corinthians 15:35-44
281 Matthew 16:26
282 Matthew 10:28
283 1 Corinthians 15:51-53
284 Venema and McKnight, *Adam and Genome*, 111-118
285 Venema and McKnight, *Adam and Genome*, 164

286 Venema and McKnight, *Adam and Genome*, 160
287 Venema and McKnight, *Adam and Genome*, 158
288 Venema and McKnight, *Adam and Genome*, xi
289 Genesis 5:5
290 Venema and McKnight, *Adam and Genome*, 174
291 Venema and McKnight, *Adam and Genome*, 146
292 Job 1:8
293 Ezekiel 14:14; James 5:11; Jude 1:14
294 Venema and McKnight *Adam and Genome*, 176
295 Venema and McKnight, *Adam and Genome*, 179
296 John 1:1-3
297 See: https://www.youtube.com/watch?v=6h0yEpqDEI8
298 Venema and McKnight, *Adam and Genome*, xi
299 Venema and McKnight, *Adam and Genome*, 30
300 Venema and McKnight, *Adam and Genome*, 53
301 Venema and McKnight, *Adam and Genome*, 53-54
302 Venema and McKnight, *Adam and Genome*, 54
303 Venema and McKnight, *Adam and Genome*, 21
304 Venema and McKnight, *Adam and Genome*, 23
305 Venema and McKnight, *Adam and Genome*, 22
306 Venema and McKnight, *Adam and Genome*, 30
307 Venema and McKnight, *Adam and Genome*, 32
308 Job 42:3
309 Venema and McKnight, *Adam and Genome*, 65
310 Venema and McKnight, *Adam and Genome*, 48
311 Venema and McKnight, *Adam and Genome*, 44
312 Venema and McKnight, *Adam and Genome*, 45
313 Venema and McKnight, *Adam and Genome*, 48
314 Huffington Post, August 1, 2013, reporting from the journal, *Science*, August 1, 2013
315 Huffington Post, August 1, 2013
316 Huffington Post, August 1, 2013
317 Venema and McKnight, *Adam and Genome*, 46
318 Genesis 9:19 NLT
319 Venema and McKnight, *Adam and Genome*, 41
320 Venema and McKnight, *Adam and Genome*, 17-18
321 Venema and McKnight, *Adam and Genome*, 18
322 Venema and McKnight, *Adam and Genome*, 19
323 Michael Denton, *Evolution: Still a Theory in Crisis* (Seattle: Discovery Institute Press, 2016), 18
324 Denton, *Still in Crisis*, 125

325 Denton, *Still in Crisis*, 114
326 Denton, *Still in Crisis*, 59
327 Denton, *Still in Crisis*, 117
328 Lewis, "The Laws of Nature," *God in the Dock* (1945), 77
329 Denton, *Still in Crisis*, 100
330 Denton, *Still in Crisis*, 87
331 Walton, John H., *The Lost World of Genesis One* (Downers Grove, IL: InterVarsity Press, 2009), 58
332 Walton, *Lost World*, 161-162
333 Walton, *Lost World*, 161-162
334 Walton, *Lost World*, 166
335 Walton, *Lost World*, 9
336 Walton, *Lost World*, 138
337 Walton, *Lost World*, 67
338 Walton, *Lost World*, 65
339 Walton, *Lost World*, 164
340 Walton, *Lost World*, 165
341 Walton, *Lost World*, 131-132
342 Walton, *Lost World*, 112
343 Walton, *Lost World*, 158
344 Walton, *Lost World*, 163
345 Walton, *Lost World*, 158
346 C.S. Lewis, "The Funeral of a Great Myth," 87-88
347 C.S. Lewis, "Evolutionary Hymn," (letter to Dorothy Sayers, 1954), in Lewis, *Poems*, 55
348 See John G. West, ed., *The Magician's Twin—C.S. Lewis on Science, Scientism, and Society* (Seattle: Discovery Institute Press, 2012), 134
349 See, West, *Magician's Twin*, 134-138
350 C.S. Lewis, *Mere Christianity* 35
351 C.S. Lewis to Bernard Acworth, Sept. 13, 1951, *Collected Letters*, vol. III, 138
352 West, *Magician's Twin*, 138
353 See, West, *Magician's Twin*, 137
354 C.S. Lewis, *Collected Letters*, 138
355 Jonathan Wynne-Jones, "Charles Darwin to receive apology from the Church of England for rejecting evolution," *The Daily Telegraph*, September 13, 2008.